Triumph 650 & 750 2-Valve Twins Owners Workshop Manual

by Jeff Clew
With an additional Chapter on the 1975 to 1983 models
by Chris Rogers

Models covered:

Model		Displacement	Years	Model		Displacement	Years
T6 Thunderbird		649cc.	'63 - '66	T140D Bonneville			
TR6 Trophy		649cc.	'63 - '71	Special		744cc.	'79 - '83
TR6R Tiger		649cc.	'70 - '73	T140ES Bonneville			
T120 Bonneville		649cc.	'63 - '75	Electro		744cc.	'80 - '83
TR65 Thunderbird		649cc.	'82 - '83	T140EX Bonneville			
TR65T Tiger Trail		649cc.	'82 only	Executive		744cc.	'81 - '83
T140V Bonneville		744cc.	'73 - '78	T140LE Bonneville			
T140V Bonneville				Royal		744cc.	'81 only
Jubilee		744cc.	'77 only	TR7V Tiger		744cc.	'73 - '78
T140E Bonneville		744cc.	'78 - '83	TR7 Tiger		744cc.	'78 - '81
TSX 2-valve		744cc.	'82 - '83	TR7T Tiger Trail		744cc.	'81 - '82

ISBN 978 0 85696 890 7

ABCDE
FGHIJ
KLMNO
PQRS
4

Printed in Malaysia   (122-6P10)

Haynes Publishing
Sparkford, Yeovil, Somerset
BA22 7JJ, England

Haynes North America, Inc
859 Lawrence Drive, Newbury Park,
California 91320, USA

Acknowledgements

Thanks are due to Triumph Motorcycles (Meriden) Limited for the supply of technical information; to the Avon Rubber Company, who kindly supplied information and technical assistance on tyre fitting; NGK Spark Plugs (UK) Ltd for information on spark plug maintenance and electrode conditions, and Renold Ltd for advice on chain care and renewal. Amal Limited supplied some of the illustrations which accompany the text of this Manual.

Special thanks are due to Tony Mead of Atkins Motorcycles, Taunton, Somerset, for allowing us the use of his showroom to photograph the machine shown on the front cover of this Manual and the use of his grounds to photograph some of the machines shown in the text.

Thanks are also due to Charlies of Bristol for supplying technical information and to M R Holland Limited of Spalding, Lincs, for information relating to rear suspension units.

About this manual

The author of this manual has the conviction that the only way in which a meaningful and easy-to-follow text can be written is to carry out the work himself, under conditions similar to those found in the average household. As a result, the hands seen in the photographs are those of the author. Even the machines are not new: examples which have covered a considerable mileage are selected, so that the conditions encountered would be typical of those encountered by the average rider/owner. Unless specially mentioned, and therefore considered essential, Triumph service tools have not been used. There are invariably alternative means of slackening or removing some vital component when service tools are not available, but risk of damage is to be avoided at all costs.

Each of the eight chapters is divided into numbered sections. Within the sections are numbered paragraphs. Cross-reference throughout the manual is quite straightforward and logical. For example, when reference is made 'See Section 6.2' it means section 6, paragraph 2 in the same chapter. If another chapter were meant, the reference would read 'See Chapter 2, section 6.2'. All photographs are captioned with a section/paragraph number to which they refer, and are always relevant to the chapter text adjacent.

Figure numbers (usually line illustrations) appear in numerical order, within a given chapter. Fig 1.1 therefore refers to the first figure in Chapter 1. Left hand and right hand descriptions of the machines and their component parts refer to the left and right when the rider is seated, facing forward.

Motorcycle manufacturers continually make changes to specifications and recommendations, and these, when notified, are incorporated into our manuals at the earliest opportunity.

Contents

1968 650 Bonneville

1972 650 Bonneville

**1972 650 Engine in duplex frame -
Primary side**

1972 650 Bonneville - Gearshift side

Introduction to the Triumph 650/750cc unit construction vertical twins

Contrary to popular belief, the first Triumph vertical twin engine was designed and manufactured as far back as 1914, although a complete machine was not built. It was not until 1933 that another twin cylinder design emerged, this time in the form of a unit-construction 650 cc engine with geared primary drive, designed by Val Page. One of the completed models, with sidecar attached, covered 500 miles in 500 minutes at Brooklands immediately after the outfit had competed successfully in that year's International Six Days Trial. This feat, which was observed by the Auto-Cycle Union throughout, won the Maudes Trophy for Triumph Motors.

In January 1936 the Triumph Engineering Company Limited was formed to take over the motor cycle manufacturing activities of Triumph Motors. A new designer, Edward Turner, was appointed and it was Turner who inspired the Speed Twin model that had sensational impact on the motor cycle world during 1937. Indeed, this is the model that is the true ancestor of today's vertical twin designs and the one that established an entirely new trend in motor cycling.

The first of the modern 650 cc models was the Thunderbird, first manufactured during 1949. Although by no means a sluggard, there was evidence of demand for a model with even higher performance and in 1953 the T110 model was announced, the coding relating to the anticipated performance in miles per hour. The T110 was one of the first models to feature swinging arm rear suspension; the earlier Thunderbird models were supplied with either a rigid frame or a sprung hub of Triumph manufacture - another Triumph innovation.

In due course the T110 was supplemented by another model capable of even higher performance, the T120 of 1958. This latter machine was named the Bonneville in recognition of Johnny Allen's record breaking attempts in America with an unsupercharged 650 cc engine encased in a cigar-like shell. The attempts took place on the Bonneville Salt Flats on 25th September, 1955 when Allen achieved a mean speed of 193.72 mph over the flying kilometre and 192.30 mph over the measured mile. These were the highest speeds ever covered on a motor cycle at that time and it is unfortunate that this remarkable achievement was somewhat clouded by squabbles at International level about the official recognition of this feat.

Production of the Bonneville model continued virtually unchanged until the end of 1962, when it was announced that the 1963 models would be built on the unit-construction principle, to bring them in line with the other models then in the current range. Further design changes, including a new frame and forks were made in 1971. It is basically this latter design that continues today and is dealt with in this manual.

The 750 cc models first appeared at the beginning of 1973 and can be regarded as 'stretched' versions of the well-proven 650 cc designs. Two models of this capacity are available, the TR7V Tiger 750 and the T140V Bonneville.

Mention should also be made of the Trophy model which started life in 1948 as a competition model, for trials and scrambles use. Early models were of 500 cc capacity only and used a development of the square-finned generator engine that was manufactured during the war. In 1951 this engine was superseded by a new close-pitch finned engine of all alloy construction, used also in modified form for the Tiger 100 model. A 650 cc Trophy model was added to the range in 1956 and eventually the 500 cc model was phased out. Production of the 650 cc Trophy model finally ceased during 1972.

Guide to machine identification

The engine number is stamped into a raised pad on the left hand crankcase, immediately below the cylinder barrel. From mid-1969 (engine no. DU86965 on) a series of Triumph motifs were rolled into the pad to render obvious any attempt at altering the number. The frame number is stamped into the left hand side of the steering head lug. Note that the engine and frame numbers should be the same.

Identification – early models

From 1963 to mid-1969 engine and frame numbers started with the DU prefix, running from DU101 to DU90282. The model designation was added to indicate the machine's actual specification, eg T120.

Identification – later models

From mid-1969 (JC00101), the system of numbering changed and a prefix was added to indicate the month and year of manufacture.

The first letter indicates the month of manufacture as follows:

A	January	H	July
B	February	J	August
C	March	K	September
D	April	N	October
E	May	P	November
G	June	X	December

The second letter indicates the year of manufacture as follows:

C	1969	J	1974	A	1979
D	1970	K	1975	B	1980
E	1971	N	1976	DA	1981
G	1972	P	1977	DA	1982
H	1973	X	1978	EA	1983

The third section is a numerical block of five figures which commences with engine number 00101. The fourth section indicates the machine specification.

Note that models are identified at all times in this manual by their Triumph year of production; this may not be the same as a machine's date of sale.

Ordering spare parts

When ordering spare parts for any of the Triumph unit-construction vertical twins, it is advisable to deal direct with an official Triumph agent who will be able to supply many of the items ex-stock. Parts cannot be obtained direst from the Triumph Engineering Company Limited; all orders must be routed through an approved agent, even if the parts required are not held in stock.

Always quote the engine and frame numbers in full. Include any letters before or after the number itself.

Use only parts of genuine Triumph manufacture. Pattern parts are available but in many instances they will have an adverse effect on performance and/or reliability. Some complete units are available on a 'service exchange' basis, affording an economic method of repair without having to wait for parts to be reconditioned. Details of the parts available, which include petrol tanks, front forks, front and rear frames, clutch plates, brake shoes etc can be obtained from any Triumph agent. It follows that the parts to be exchanged must be acceptable before factory reconditioned replacements can be supplied.

Some of the more expendable parts such as spark plugs, bulbs, tyres, oils and greases etc, can be obtained from accessory shops and motor factors, who have convenient opening hours, charge lower prices and can often be found not far from home. It is also possible to obtain parts on a Mail Order basis from a number of specialists who advertise regularly in the motor cycle magazines.

ENGINE NUMBER LOCATION (650 CC TWIN)

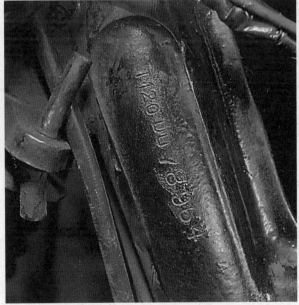

FRAME NUMBER LOCATION (650 CC TWIN)

Routine maintenance

Periodic routine maintenance is a continuous process that commences immediately the machine is used. It must be carried out at specified mileage recordings or on a calendar date basis if the machine is not used regularly, whichever falls soonest. Maintenance should be regarded as an insurance policy, to help keep the machine in peak condition and to ensure long, trouble-free service. It has the additional benefit of giving early warning to any faults that may develop and will act as a regular safety check, to the obvious advantage of both rider and machine alike.

The various maintenance tasks are described under their respective mileage and calendar headings. Accompanying diagrams are provided, where necessary. It should be remembered that the interval between the various maintenance tasks serves only as a guide. As the machine gets older or is used under particularly adverse conditions, it would be advisable to reduce the period between each check.

Some of the tasks are described in detail, where they are not mentioned fully as a routine maintenance item in the text. If a specific item is mentioned, but not described in detail, it will be covered fully in the appropriate chapter. No special tools are required for the normal routine maintenance tasks. The tools contained in the tool kit supplied with every new machine will prove adequate for each task or if they are not available, the tools found in the average household will usually suffice.

RM1. Engine oil must be changed every 3,000 miles

Weekly or every 250 miles (400 km)

Check level in oil tank or reservoir and top up if necessary.
Check level in primary chaincase and lubricate the rear chain.
Check battery acid level and tyre pressures.

Monthly or every 1000 miles (1600 km)

Change oil in primary chaincase.
Lubricate the control cables (see accompanying diagram) and grease the swinging arm fork pivot (grease nipple on underside).
Remove, clean and re-lubricate the final drive chain.
Check all nuts, bolts etc for tightness.
Adjust tension of primary and final drive chains.

Six weekly or every 1500 miles (2400 km)

Apply light smear of grease to contact breaker cam.
Check adjustment of steering head bearings.
Change engine oil and clean filters in lubricating system.

RM2. Battery is often the most neglected of all components

Three monthly or every 3000 miles (4800 km)

Change the engine oil and clean the filters in the lubrication system.
Lubricate the contact breaker.

Six monthly or every 6000 miles (9600 km)

Check oil level in gearbox.
Examine front forks for oil leakage.
Grease brake pedal spindle.
Check valve clearances and ignition timing.
Clean and adjust spark plugs.
Clean air filter and carburettor(s).
Adjust contact breaker.

Yearly or every 12,000 miles (19,200 km)

Grease wheel and steering head bearings.

These latter two tasks will necessitate a certain amount of dismantling, details of which are given in Chapters 5 and 6.

It should be noted that even when six monthly and yearly maintenance tasks have to be undertaken, the weekly, monthly, six weekly and three monthly services must be completed. There is no stage at any point during the life of the machine when a routine maintenance task can be ignored.

Routine maintenance and capacities data

Engine oil capacity:
Oil tank models ... ...   5 Imp pints (2.84 litre)
Oil in frame models ... ...   4 Imp pints (2.27 litre)

Primary chaincase:
All models 1963-69 ... ...   5/8 Imp pint (350 cc)
All models 1970 on ... ...   ¼ Imp pint (150 cc)

Gearbox ... ...   7/8 Imp pint (500 cc)

Front forks:
Engine nos DU101-DU5824
(1963 models only) ... ...   ¼ Imp pint (150 cc)
All other models ... ...   1/3 Imp pint (190 cc)

Contact breaker gap ... ...   0.014-0.016 in (0.35-0.40 mm)

Spark plug gap:
Engine nos DU101-DU66245
(1963-67 models fitted with 4CA
contact breaker assembly)... ...   0.020 in (0.50 mm)
Engine no DU66246 onwards
(all models fitted with 6CA or
10CA contact breaker assembly) 0.025 in (0.64 mm)

Tyre pressures:
All models 1963-67 ... ...   20 psi front, 18 psi rear
All other models ... ...   24 psi front, 25 psi rear

Mention has not been made of the lighting equipment, horn and speedometer, all of which must be checked frequently to ensure they are in good working order. The tyres also should be checked regularly and renewed if they wear unevenly, have splits or cracks in the side walls or if the depth of tread approaches the statutory minimum. Neglect of any of these points may render the owner liable to prosecution, apart from creating a safety hazard.

RM3. Check chain case oil content regularly and top up when necessary

RM4. Do not neglect to check oil level in gearbox. Low oil content causes rapid wear

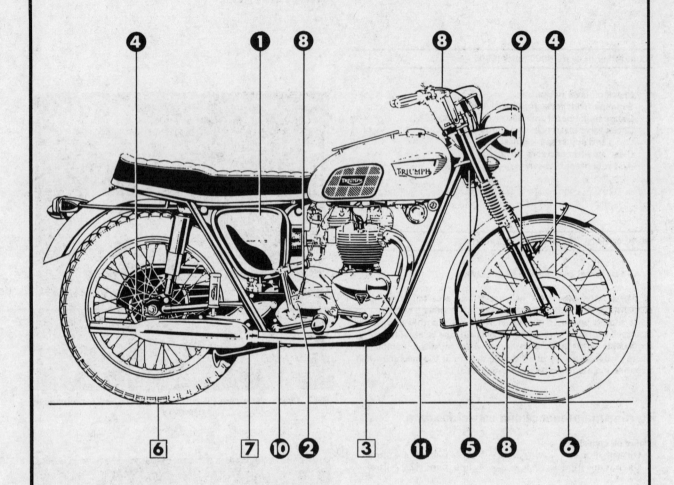

FIG.RM5. LUBRICATION POINTS

See page 11 for Key

Guide to lubrication

Component	Key on illustration
Engine oil tank ...	1
Gearbox ...	2
Primary chaincase ...	3
Wheel hubs ...	4
Steering head ...	5
Brake cam spindle ...	6
Brake pedal spindle ...	7
Exposed cables ...	8
Telescopic fork ...	9
Swinging fork pivot ...	10
Contact breaker cam ...	11
All brake rod joints and pins ...	—

Recommended lubricants

Component	Type of lubricant	Castrol Grade
Engine ...	Summer: 20W/50 Multigrade ...	Castrol GTX
	Winter: 10W/30 Multigrade ...	Castrolite
Gearbox ...	SAE 50 Monograde ...	Castrol Grand Prix
		Castrol Hypoy EP 90 (750 cc models)
Primary chaincase ...	10W/30 Multigrade ...	Castrolite
All grease points: Cam spindles etc. ...	Lithium based high melting point ...	Castrol LM Grease
All lubrication points; including front forks and cables ...	Light oil, 10W/30 Multigrade 20W/50 Multigrade ...	Castrolite Castrol GTX (750 cc models)
Hydraulic disc brake ...	High quality brake fluid ...	Castrol Girling Universal Brake and Clutch Fluid

Safety first!

Professional motor mechanics are trained in safe working procedures. However enthusiastic you may be about getting on with the job in hand, do take the time to ensure that your safety is not put at risk. A moment's lack of attention can result in an accident, as can failure to observe certain elementary precautions.

There will always be new ways of having accidents, and the following points do not pretend to be a comprehensive list of all dangers; they are intended rather to make you aware of the risks and to encourage a safety-conscious approach to all work you carry out on your vehicle.

Essential DOs and DON'Ts

DON'T start the engine without first ascertaining that the transmission is in neutral.

DON'T suddenly remove the filler cap from a hot cooling system – cover it with a cloth and release the pressure gradually first, or you may get scalded by escaping coolant.

DON'T attempt to drain oil until you are sure it has cooled sufficiently to avoid scalding you.

DON'T grasp any part of the engine, exhaust or silencer without first ascertaining that it is sufficiently cool to avoid burning you.

DON'T allow brake fluid or antifreeze to contact the machine's paintwork or plastic components.

DON'T syphon toxic liquids such as fuel, brake fluid or antifreeze by mouth, or allow them to remain on your skin.

DON'T inhale dust – it may be injurious to health (see *Asbestos* heading).

DON'T allow any spilt oil or grease to remain on the floor – wipe it up straight away, before someone slips on it.

DON'T use ill-fitting spanners or other tools which may slip and cause injury.

DON'T attempt to lift a heavy component which may be beyond your capability – get assistance.

DON'T rush to finish a job, or take unverified short cuts.

DON'T allow children or animals in or around an unattended vehicle.

DON'T inflate a tyre to a pressure above the recommended maximum. Apart from overstressing the carcase and wheel rim, in extreme cases the tyre may blow off forcibly.

DO ensure that the machine is supported securely at all times. This is especially important when the machine is blocked up to aid wheel or fork removal.

DO take care when attempting to slacken a stubborn nut or bolt. It is generally better to pull on a spanner, rather than push, so that if slippage occurs you fall away from the machine rather than on to it.

DO wear eye protection when using power tools such as drill, sander, bench grinder etc.

DO use a barrier cream on your hands prior to undertaking dirty jobs – it will protect your skin from infection as well as making the dirt easier to remove afterwards; but make sure your hands aren't left slippery. Note that long-term contact with used engine oil can be a health hazard.

DO keep loose clothing (cuffs, tie etc) and long hair well out of the way of moving mechanical parts.

DO remove rings, wristwatch etc, before working on the vehicle – especially the electrical system.

DO keep your work area tidy – it is only too easy to fall over articles left lying around.

DO exercise caution when compressing springs for removal or installation. Ensure that the tension is applied and released in a controlled manner, using suitable tools which preclude the possibility of the spring escaping violently.

DO ensure that any lifting tackle used has a safe working load rating adequate for the job.

DO get someone to check periodically that all is well, when working alone on the vehicle.

DO carry out work in a logical sequence and check that everything is correctly assembled and tightened afterwards.

DO remember that your vehicle's safety affects that of yourself and others. If in doubt on any point, get specialist advice.

IF, in spite of following these precautions, you are unfortunate enough to injure yourself, seek medical attention as soon as possible.

Asbestos

Certain friction, insulating, sealing, and other products – such as brake linings, clutch linings, gaskets, etc – contain asbestos. *Extreme care must be taken to avoid inhalation of dust from such products since it is hazardous to health.* If in doubt, assume that they *do* contain asbestos.

Fire

Remember at all times that petrol (gasoline) is highly flammable. Never smoke, or have any kind of naked flame around, when working on the vehicle. But the risk does not end there – a spark caused by an electrical short-circuit, by two metal surfaces contacting each other, by careless use of tools, or even by static electricity built up in your body under certain conditions, can ignite petrol vapour, which in a confined space is highly explosive.

Always disconnect the battery earth (ground) terminal before working on any part of the fuel or electrical system, and never risk spilling fuel on to a hot engine or exhaust.

It is recommended that a fire extinguisher of a type suitable for fuel and electrical fires is kept handy in the garage or workplace at all times. Never try to extinguish a fuel or electrical fire with water.

Note: *Any reference to a 'torch' appearing in this manual should always be taken to mean a hand-held battery-operated electric lamp or flashlight. It does **not** mean a welding/gas torch or blowlamp.*

Fumes

Certain fumes are highly toxic and can quickly cause unconsciousness and even death if inhaled to any extent. Petrol (gasoline) vapour comes into this category, as do the vapours from certain solvents such as trichloroethylene. Any draining or pouring of such volatile fluids should be done in a well ventilated area.

When using cleaning fluids and solvents, read the instructions carefully. Never use materials from unmarked containers – they may give off poisonous vapours.

Never run the engine of a motor vehicle in an enclosed space such as a garage. Exhaust fumes contain carbon monoxide which is extremely poisonous; if you need to run the engine, always do so in the open air or at least have the rear of the vehicle outside the workplace.

The battery

Never cause a spark, or allow a naked light, near the vehicle's battery. It will normally be giving off a certain amount of hydrogen gas, which is highly explosive.

Always disconnect the battery earth (ground) terminal before working on the fuel or electrical systems.

If possible, loosen the filler plugs or cover when charging the battery from an external source. Do not charge at an excessive rate or the battery may burst.

Take care when topping up and when carrying the battery. The acid electrolyte, even when diluted, is very corrosive and should not be allowed to contact the eyes or skin.

If you ever need to prepare electrolyte yourself, always add the acid slowly to the water, and never the other way round. Protect against splashes by wearing rubber gloves and goggles.

Mains electricity and electrical equipment

When using an electric power tool, inspection light etc, always ensure that the appliance is correctly connected to its plug and that, where necessary, it is properly earthed (grounded). Do not use such appliances in damp conditions and, again, beware of creating a spark or applying excessive heat in the vicinity of fuel or fuel vapour. Also ensure that the appliances meet the relevant national safety standards.

Ignition HT voltage

A severe electric shock can result from touching certain parts of the ignition system, such as the HT leads, when the engine is running or being cranked, particularly if components are damp or the insulation is defective. Where an electronic ignition system is fitted, the HT voltage is much higher and could prove fatal.

Chapter 1 Engine

Contents

Specifications

The Triumph 650/750 cc unit-construction vertical twins all employ the same basic engine/gear unit in which the gearbox is an integral part of the engine assembly. The same dismantling and reassembly procedure is applicable to all the 650/750 cc unit-construction models.

Engine (T120 Bonneville, TR6 Trophy, TR7V Tiger 750, T140V Bonneville)

Type	Twin cylinder, vertically mounted	
Cylinder head	Aluminium alloy	
Cylinder barrel	Cast iron	
Bore	71 mm	76 mm
Stroke	82 mm	82 mm
Capacity	649 cc	744 cc ^{x}
Power output (bhp)	47 @ 6700 rpm *	49 @ 6200 rpm $^+$
Compression ratio	9 : 1	8.5 : 1

     * T120 Bonneville
     $^+$ TR7V Tiger 750
     ^{x} Early T140V models 724 cc (75 x 82 mm)

Crankshaft

Main bearing - left hand (drive) side	Single lipped roller bearing
Size	$2^{13/16}$ in x $1^{1/8}$ in x $^{13/16}$ in
Main bearing - right hand (timing) side	Ball journal
Size	$2^{13/16}$ in x $1^{1/8}$ in x $^{13/16}$ in
Big end journal diameter	1.6235 in - 1.6240 in
Minimum regrind diameter...	1.6035 in - 1.6040 in
Permisssible end float	0.003 in - 0.017 in

Connecting rods
Length (centres)	6.499 in - 6.501 in	6.001 in - 5.999 in (1140V)
Big end bearings - type	Steel backed white metal	} As 650 cc models
Bearing side clearance	0.012 in - 0.016 in	

Oil pressure
Normal running	65 - 80 psi	} As 650 cc models
Idling	20 - 25 psi	

Camshafts
Journal bearing diameter: left hand	0.8100 in - 0.8105 in	} As 650 cc models
right hand	0.8730 in - 0.8735 in	
End float	0.013 in - 0.020 in	
Cam lift: inlet and exhaust	0.314 in	0.347 in and 0.305 in
Base circle diameter	0.812 in	(inlet and exhaust)

Tappets
Clearance (cold) inlet	0.002 in	0.008 in
exhaust	0.004 in	0.006 in

Pistons
Clearance (top of skirt)	0.0106 in - 0.0085 in	} As 650 cc models
(bottom of skirt)	0.0061 in - 0.0046 in	
Oversizes available	+ 0.010 in, + 0.020 in, + 0.040 in	

Piston rings
Compression (two rings, tapered)
Width	0.0615/0.0625 in	0.121 in - 0.113 in
Radial depth	0.092/0.100 in	0.0625 in - 0.0615 in
End gap	0.010/0.014 in	0.008 in - 0.013 in

Oil control ring
Width	0.124/0.125 in	0.121 in
Radial depth	0.092/0.100 in	0.125 in
End gap	0.010/0.014 in	0.010 - 0.040 in

Valves
Stem diameter: inlet	0.3095/0.3100 in	} As 650 cc models
exhaust	0.3090/0.3095 in	
Head diameter: inlet	1.592/1.596 in	
exhaust	1.434/1.440 in	

Valve guides
Material	Aluminium - bronze	} As 650 models
Bore diameter: inlet and exhaust	0.3127/0.3137 in	
Outside diameter: inlet and exhaust	0.5005/0.5010 in	
Length: inlet	1.969 in	
exhaust	2.172 in	

Valve springs
Free length: outer	1½ in Green spot	} As 650 cc models
inner	1 17/32 in Red Spot	

Torque wrench settings
Flywheel bolts	33 lb ft	As 650 cc models
Connecting rod bolts	28 lb ft	22 lb ft
Crankcase bolts	13 lb ft	
Crankcase studs	20 lb ft	
Cylinder base nuts	35 lb ft	
Cylinder head bolts (3/8 in diam)	18 lb ft *	
Cylinder head bolts (5/16 in diam)	15 lb ft	As 650 cc models
Rocker box nuts and bolts	5 lb ft	
Rocker spindle domed nuts	22 lb ft	
Oil pump nuts	5 lb ft	
Rotor retaining nut	30 lb ft	40
Stator retaining nuts	20 lb ft	As 650 cc models

* Reduction from original recommendation of 25 lb ft

Valve timing
Inlet valve opens	34° BTDC	0.190 in @ TDC
Inlet valve closes	55° ABDC	—
Exhaust valve opens	55° BBDC	0.130 in @ TDC
Exhaust valve closes	34° ATDC	

Note: All valve clearances must be set at 0.020 in whilst checking

Data on earlier unit-construction engines (prior to engine No DU 66246) (6T Thunderbird, TR6 Trophy and T120 Bonneville models)

Data given if different from that in preceding section

Compression ratio ... ...
7.5 : 1 6T models up to engine no DU 44393
8.5 : 1 TR6 models up to engine no DU 44393
8.5 : 1 T120 models up to engine no DU 24874

Power output (bhp) ... ...
37 @ 6700 rpm 6T models as above
40 @ 6500 rpm TR6 models as above
46 @ 6500 rpm T120 models as above

Pistons
Clearance (top of skirt) ... ...
0.0088 in - 0.0098 in 6T models DU 101 - DU 5824 and TR6
0.0046 in - 0.0057 in 6T models DU 5825 - DU 44393
0.0093 in - 0.0103 in T120 and T120TT

(bottom of skirt) ... ...
0.0033 in - 0.0043 in 6T models DU 101 - DU 5824 and TR6
0.0016 in - 0.0027 in 6T models DU 5825 - DU 44393
0.0038 in - 0.0048 in T120
0.0073 in - 0.0083 in T120TT

Valves
Head diameter: inlet ... ...
1.5 in 6T and TR6 DU 101 - DU 44393
1.5 in T120 DU 101 - DU 5824
1.594 in T120 DU 5825 onward

exhaust ... ...
1.344 in 6T and TR6 DU 101 - DU 44393
1.344 in T120 DU 101 - DU 5824
1.229 in T120 DU 5825 onward

Valve springs
Free length: outer ... ...
$1^{5}/8$ in 6T DU 101 onward)
TR6 DU 101 onward) Red spot
T120 DU 24875 DU 44393)
$2^{1}/32$ in T120 DU 101 - DU 24874 White spot

inner ... ...
$1^{17}/32$ in 6T DU 101 onward)
TR6 DU 101 onward) Red spot
T120 DU 24875 - DU 44393)
$1^{5}/8$ in T120 DU 101 - DU 24874 White spot

Valve timing
Inlet valve opens ... ...
25° BTDC 6T models, DU 101 onward
34° BTDC TR6 models, DU 101 - DU 44393 and T120 models, DU 101 - DU 24874

Inlet valve closes ... ...
52° ABDC 6T models, DU 101 onward
55° ABDC TR6 models, DU 101 - DU 44393 and T120 models, DU 101 - DU 24874

Exhaust valve opens ... ...
60° BBDC 6T models, DU 101 onward
55° BBDC TR6 models, DU 101 - DU 44393 and T120 models, DU 101 - DU 24874

Exhaust valve closes ... ...
17° ABDC 6T models, DU 101 onward
27° ABDC TR6 models, DU 101 - DU 44393 and T120 models, DU 101 - DU 24874

Note: All valve clearances must be set at 0.020 in whilst checking

Data on USA only export model T120TT Bonneville TT Special

Data given if different from that in preceding sections

Compression ratio T120TT ... ...
11 : 1

Power output (bhp) ... ...,. ... ...
54 @ 6500

This model was discontinued after engine No DU 66245

1 General description

The engine fitted to the Triumph unit-construction vertical twins is of the combined engine and gearbox type, in which the gearbox casting forms an integral part of the right hand crankcase and the primary chaincase, an integral part of the left hand chaincase. Aluminium alloy is used for all of the engine castings, with the exception of the cylinder barrel which is of cast iron.

The cylinder head has cast-in austenitic valve seat inserts and houses the overhead valves which are actuated by rocker arms enclosed within the detachable rocker boxes. The pushrods are of aluminium alloy with hardened end pieces.

'H' section connecting rods of hinduminium alloy, with detachable caps and steel-backed shell bearings, carry aluminium alloy die-cast pistons, each with two compression rings and one oil scraper ring. The two-throw crankshaft has a detachable shrunk-on cast iron flywheel, retained in a central position by three high tensile steel bolts. The cast iron cylinder barrel houses the press-fit tappet guide blocks.

The separate inlet and exhaust camshafts operate in sintered bronze bushes, mounted transversely in the upper part of the crankcase. The camshafts are driven by the train of timing gears, from the right hand end of the crankshaft. The inlet camshaft provides the drive for the oil pump and the rotary breather valve disc, whilst the exhaust camshaft drives the contact breaker and, on some models, the tachometer drive gearbox.

Power from the engine is transmitted in the conventional manner through the engine sprocket and primary chain to the clutch unit that incorporates a shock absorber. Chain tension is controlled by an adjustable chain tensioner, immersed within the oil content of the chaincase.

2 Operations with engine in frame

It is not necessary to remove the engine unit from the frame unless the crankshaft assembly and/or main bearings require attention. Most operations can be accomplished with the engine in place, such as:
1   Removal and replacement of cylinder head.
2   Removal and replacement of cylinder barrel and pistons.
3   Removal and replacement of alternator.
4   Removal and replacement of primary drive components.
5   Removal and replacement of oil pump and contact breaker assembly.
When several operations need to be undertaken simultaneously, for example during a major overhaul, it will probably be advantageous to remove the complete engine unit from the frame, an operation that should take approximately 2 to 2½ hours. This is a two man job; the engine unit weighs 135 lbs.

3 Operations with engine removed

1   Removal and replacement of the main bearings.
2   Removal and replacement of the crankshaft assembly.
3   Removal and replacement of the camshafts.
4   Renewing the camshaft and timing gear bushes.

4 Method of engine/gearbox removal

As described previously, the engine and gearbox are built in unit, although it is not necessary to remove the unit complete in order to gain access to either component unless a major overhaul is contemplated, or more working space is required. It is not possible to dismantle the engine fully, until the engine/gearbox unit has been removed from the frame and refitting cannot take place until the crankcases have been reassembled. The crankcases cannot be separated unless both the outer and inner gearbox end covers are removed too, exposing the gear clusters.

5 Removing the engine/gearbox unit

1   Place the machine on the centre stand and make sure it is standing firmly, on level ground.
2   Turn off the petrol taps and disconnect the petrol pipes by either unscrewing the union nut joints or by pulling the pipes from their push-fit connections at the bottom of the float chamber(s).
3   Remove the petrol tank. This is secured by two nuts or bolts underneath the nose, which may be wired together, and by a bolt that passes through a lug welded to the rear of the tank. Some models exported to the USA have reflectors fitted below the front of the petrol tank, secured by the front mounting nuts. The reflector units must be removed first. If necessary, the reflector lenses can be prised out of position to gain better access to the nuts or bolts. On later models the reflectors are attached to a bar across the frame tubes and need not be removed. Make sure that the rubbers isolating the tank from vibration are not misplaced and lost.
4   On some of the earlier models it will be necessary to remove the screws retaining the rear of the headlamp nacelle in order to gain sufficient clearance.
5   Raise the dual seat and detach the battery leads, or if a fuse is fitted, remove the fuse from the holder.
6   Detach the spark plug leads and remove the top bolts from the clamps securing the twin ignition coils (one on each side of the top frame tube) and turn the coils so that access is available to the cylinder head torque stays. Great care must be taken not to damage the alloy casing of the coils since this may lead to premature failure. If there is any risk of accidental damage, remove the coils. Snap connectors make electrical disconnection easy. On post 1970 models they are in a different location.
7   Unscrew the nuts retaining the torque stays to the cylinder head, remove the front and rear torque stay mounting bolts, together with their distance pieces, then detach the torque stays.
8   Remove the two domed nuts on the left hand side of the rocker boxes and draw off the rocker oil feed pipe. The pipe should be steadied during this operation; if it bends to an acute angle there is risk of fracture.
9   Although it is just possible to remove the engine on some models, with the rocker boxes attached, the amount of clearance between the top tube of the frame is minimal. Since the rocker boxes will have to be removed at a later stage, they should be detached now, to obviate the risk of damage during engine removal and to make the engine lift out much easier. Lift out the pushrods, one by one, marking each so that it is replaced in the same position during reassembly.
10 Detach the top of the carburettor(s) and withdraw the slide and needle assembly. It is best to tape these components to a frame member, so that they are well clear of the engine and cannot be damaged during its removal.
11 Detach the carburettor body(s) complete by removing the retaining nuts and washers from the inlet manifold. Place them in a safe place, away from the working area. On some models it may be necessary to remove the air cleaner first.
12 Disconnect the speedometer cable from the speedometer head (underside) and the tachometer cable (if fitted) from the drive gearbox found to the front of the left hand crankcase. On models fitted with the nacelle type headlamp, it will be necessary to remove the headlamp front before the speedometer cable can be disconnected. The headlamp rim is released when the screw adjacent to the speedometer head is slackened.
13 Place a receptacle of 6 pints capacity below the oil tank drain plug and remove the plug or, on earlier models remove the main oil feed pipe. Leave the oil to drain for about 15 minutes whilst the remainder of the dismantling is taking place.
14 It is opportune at the same time to drain both the gearbox and the primary chaincase by removing their respective drain

5.2 Float chamber unions have push-on connectors

5.3 Petrol tank has three-point fixing

5.5 Battery is housed below dualseat

5.6 Spark plug leads are push fit in coils

5.7 Torque stays must be removed completely

5.8 Support pipe whilst loosening nuts to prevent fracture

5.9a Rocker boxes should be removed before lifting engine

5.9b Mark pushrods as they are withdrawn

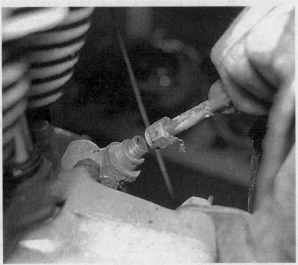
5.12 Tachometer drive is taken from end of exhaust camshaft

5.17a Exhaust pipes fit over stubs and are retained by split clamps

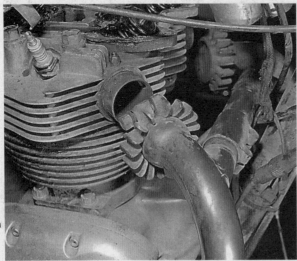
5.17b Do not omit to slacken clamps on balance pipe

5.18 Footrests are retained by nut behind engine plates

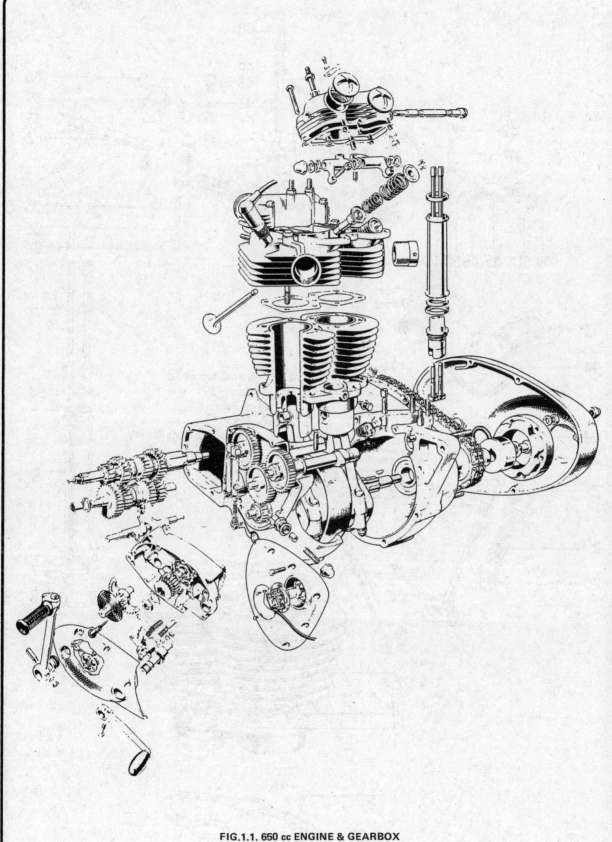

FIG.1.1. 650 cc ENGINE & GEARBOX

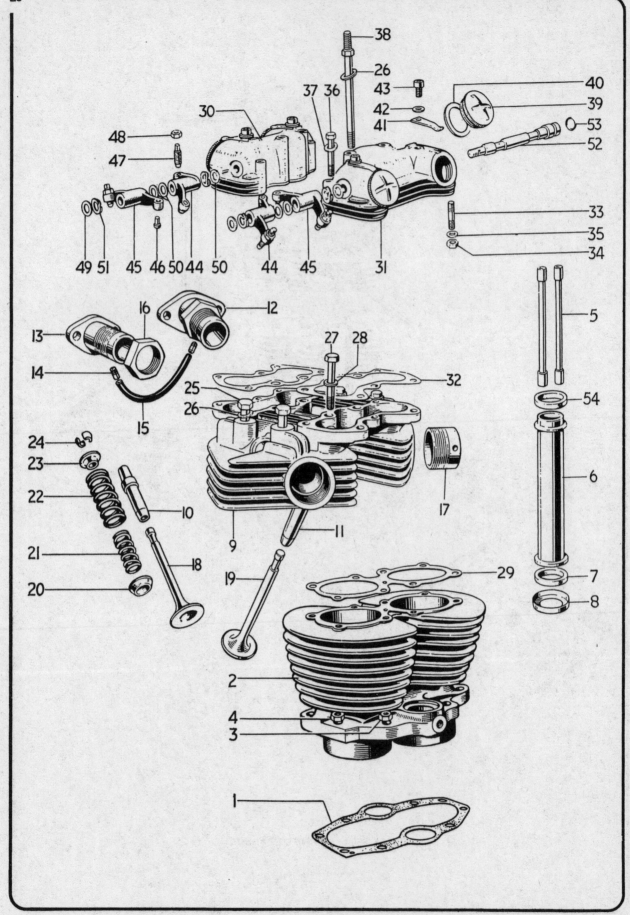

FIG.1.2 CYLINDER HEAD AND BARREL, T120 MODEL

1 Cylinder base gasket

2 Cylinder barrel

3 Inner cylinder base nut - 4 off

4 Outer cylinder base nut - 4 off

5 Push rod - 4 off

6 Push rod cover tube - 2 off

7 Rubber washer - 2 off

8 Bottom cup - 2 off

9 Cylinder head complete with valve guides

10 Inlet valve guide - 2 off

11 Exhaust valve guide - 2 off

12 Carburettor adaptor, left

13 Carburettor adaptor, right

14 Balance pipe connector - 2 off

15 Balance pipe

16 Adaptor locknut - 2 off

17 Exhaust pipe stub - 2 off

18 Inlet valve - 2 off

19 Exhaust valve - 2 off

20 Bottom cup for valve springs - 4 off

21 Inner valve spring - 4 off

22 Outer valve spring - 4 off

23 Top collar - 4 off

24 Split collets - 8 off

25 Cylinder head bolt - 4 off

26 Plain washer

27 Bolt

28 Washer

29 Cylinder head gasket

30 Inlet rocker box

31 Exhaust rocker box

32 Rocker box gaskets - 2 off

33 Rocker box stud - 6 off

34 Nut - 6 off

35 Plain washer - 6 off

36 Rocker box bolt - 4 off

37 Plain washer - 4 off

38 Cylinder head bolt for torque stay - 4 off

39 Rocker inspection cap - 4 off

40 Inspection cap washer - 4 off

41 Locking spring - 4 off

42 Fibre washer - 4 off

43 Screw - 4 off

44 Rocker arm (right exhaust, left inlet) - 2 off

45 Rocker arm (left exhaust, right inlet) - 2 off

46 Rocker ball pin - 4 off

47 Rocker adjusting pin - 4 off

48 Adjuster lock nut - 4 off

49 Thrust washer (3/8 in) - 2 off

50 Thrust washer (1/2 in) - 6 off

51 Spring washer - 4 off

52 Rocker spindle - 2 off

53 Sealing rubber - 2 off

54 Rubber washer - 2 off

plugs. The gearbox holds nearly one pint of oil and the chaincase just over a half-pint. Remove also the hexagon headed filter drain plug located underneath the engine, close to the bottom engine mounting lug. Access is from the left hand side; the filter plug is at an angle.

15 Slacken off the clutch adjuster at the end of the handle-bar control lever and, after removing the rubber cover, the adjuster at the point where the clutch cable enters the gearbox. Unscrew the inspection cap that threads into the gearbox end cover, and displace the clutch cable nipple from the end of the operating arm. Unscrew the cable adjuster from the gearbox and remove it, complete with cable.

16 Earlier models with no inspection cap are fitted with a slotted barrel-shaped connector, in which the clutch cable nipple seats. This is revealed when the clutch cable adjuster is unscrewed from the gearbox and the cable pulled upwards.

17 Slacken the finned clamps around the exhaust pipes and detach the pipes from the stays that bolt to the front engine mounting or crankcase stud. Drive the exhaust pipes off the exhaust port stubs with a rawhide mallet. Models fitted with a balance pipe between both exhaust pipe bends should have their respective clamps slackened first. Remove the complete exhaust system.

18 If the footrests are of the type that bolt direct to the lower frame tubes, beneath the engine, they must be removed. First detach the brake pedal from the left hand rear engine plate, then slacken and remove the nut that retains each footrest. This also applies to footrests attached to the rear engine plates.

19 Remove the spring link from the final drive chain and remove the chain to prevent it from falling. It is advisable to remove or tilt upwards the rear chainguard, so that it will not be damaged during engine removal.

20 Disconnect the two generator leads on the underside of the engine. Earlier models may have either 3 or 5 snap connectors.

21 Detach the rear engine plates by removing the four retaining bolts and single nut. Remove a nut and washer from the end of the front and bottom engine mounting studs and the engine unit is free within the frame.

22 If the front and lower engine mounting studs are now withdrawn, the engine can be lifted clear of the frame. This is not an easy task on account of the weight of the unit (135 lbs), either a hoist or a second pair of hands are needed. Do not misplace the distance piece fitted to the lower engine mounting stud which will fall clear after the stud is withdrawn. Withdraw the engine unit from the right of the machine.

6 Dismantling the engine - general

1 Before commencing work on the engine unit, the external surfaces should be cleaned thoroughly. A motor cycle engine has very little protection from road grit and other foreign matter, which will find its way into the engine if this simple precaution is not observed. One of the proprietary cleaning compounds such as Gunk or Jizer should be used especially if the compound is allowed to work into the film of oil and grease before it is washed away. When washing down, make sure the water cannot enter the electrical system or the now exposed inlet port(s). It will, of course, be necessary to replace the rocker boxes temporarily if this method of cleansing is employed.

2 Never use undue force to remove any stubborn part, unless mention is made of this requirement. There is invariably good reason why a part is difficult to remove, often because the dismantling operation has been tackled in the wrong sequence. Dismantling will be made easier if a simple engine stand is made up that will correspond with the engine mounting points. This arrangement will permit the complete unit to be clamped rigidly to the work bench, leaving both hands free.

7 Dismantling the engine - removing the cylinder head, barrel and pistons

1 Slacken and remove the five remaining cylinder head bolts one turn at a time until the load is released. 650 cc models have

a single bolt in the centre of the cylinder head, close to the exhaust valve rocker box. 750 cc models have two Allen screws, making a total of ten cylinder head holding down bolts. It is not possible to use conventional bolts in view of the limited amount of space available.

2 When all the bolts have been withdrawn, the cylinder head can be lifted off together with the cylinder head gasket. Unless damaged, in which case renewal is essential, the cylinder head gasket can be re-used after it has been annealed, as described later.

3 Lift off the pushrod cover tubes at the back and the front of the cylinder barrel. Discard the rubber seals, they must be renewed (if an oil tight engine is desired).

4 Turn the engine until both pistons are at top dead centre (TDC), then remove the eight nuts and washers around the base of the cylinder barrel. Before the barrel is lifted, place a rubber wedge between the inlet and exhaust valve tappets, to prevent them from falling from their guides as the cylinder barrel is raised.

5 Lift the cylinder barrel gently, taking care to support both pistons as they clear the cylinder bores. Slip rubber protectors over the now exposed cylinder base studs because they will damage the piston skirts if the pistons are permitted to drop free. Remove the cylinder base gasket and check that the two locating dowels are positioned correctly for subsequent re-assembly.

6 Before the cylinder barrel is put aside, withdraw the tappets and mark them so that they are replaced in their identical positions. Failure to observe this precaution may result in excessive tappet and cam wear and a very noisy engine.

7 Remove the wire circlips from each piston and push out the gudgeon pin, so that each piston can be detached from its connecting rod. It is probable the gudgeon pins will be a tight fit, in which case the pistons should be heated to expand the metal around the gudgeon pin boss. The alternative is to use Triumph service tool Z72 which will press each gudgeon pin out of position. Always support the connecting rod when any pressure is applied to either the piston or the gudgeon pin, otherwise there is risk of distortion.

8 Discard the circlips. They should never be re-used; new replacements are essential to eliminate the possibility of the circlips working loose whilst the engine is running and causing the gudgeon pin to make contact with the cylinder bore.

9 Mark each piston with pencil, inside the skirt, to ensure they are replaced in identical positions. If this precaution is not observed, a high rate of wear (oil consumption) may occur.

8 Dismantling the engine - removing the contact breaker and timing cover

1 The contact breaker assembly is housed within the timing cover on the right hand side of the engine, behind a cover plate retained by two crosshead screws. Remove the two screws and detach the cover plate and gasket. Remove the centre bolt that retains the automatic timing unit in position and screw in Triumph service tool D782 (engine nos from DU 66246) or D484 (engine nos up to DU 66245). This will withdraw the unit from its taper on the exhaust camshaft.

2 If the correct Triumph service tool is not available, the unit can be released by screwing a long bolt of the correct thread size ($\frac{5}{16}$ inch UNF) into the thread cut in the unit's centre. If the head of the long bolt is gripped with a Mole wrench, the taper can be broken by tapping the jaws of the wrench lightly with a hammer, as shown in the accompanying photo.

3 If the eight crosshead screws retaining the timing cover are now removed, the timing cover can be withdrawn. It may be necessary to give the cover a few light taps around the edge with a rawhide mallet until it is free. Before the cover is removed completely, detach the lead to the oil pressure warning light (post 1968 models) found beneath the rubber cover over the pressure switch attached to the forward end and the black/white and black/yellow leads from the contact breaker plate.

5.21 Detach engine plates from rear of engine

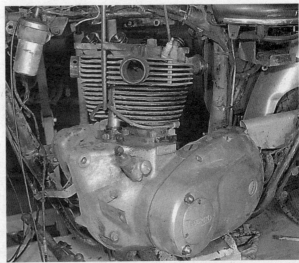

5.22a Support engine in sling whilst lifting with hoist

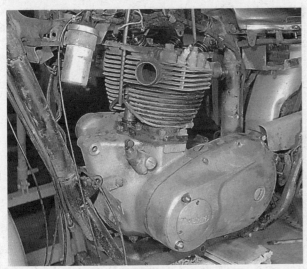

5.22b Raise front to clear front mounting then ...

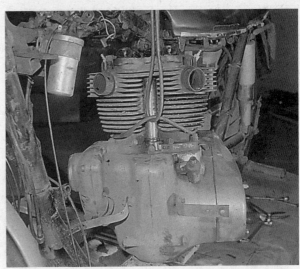

5.22c ... bring rear to right and withdraw from right hand side

7.1 650 cc models have five cylinder head bolts still to remove

7.2 Cylinder head gasket can be re-used if undamaged

7.4 Wedge tappets before lifting barrel or they may fall from guides

8.1 Contact breaker assembly is retained by two pillar bolts

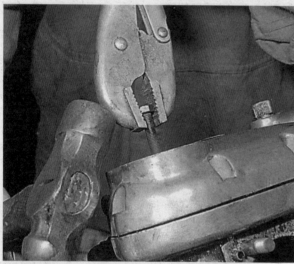

8.2a Mole wrench and bolt substitute for service tool if used carefully

8.2b Automatic timing unit is pegged to aid correct relocation

9.1 Oil pump is retained by two domed nuts

9.2. Nuts retaining camshaft pinions are LEFT HAND threaded.

9 Dismantling the engine - removing the oil pump and timing pinions

1 The oil pump, exposed after the timing cover has been removed, is retained in position by two conical nuts which, when removed, permit the pump to be drawn off the mounting studs. There is a paper gasket behind the pump which must be renewed each time the pump is removed.

2 To remove the timing pinions, first unscrew the nut that retains the crankshaft pinion to the end of the crankshaft and the nut retaining each camshaft pinion to its camshaft. Note that the camshaft pinion retaining nuts each have a LEFT HAND thread, whereas the crankshaft pinion retaining nut has a normal right hand thread. It is not necessary to remove the camshaft pinions for the crankcases to be separated, unless attention to the camshafts is required. It is preferable to leave them in situ, with their retaining nuts in position since both pinions are very difficult to extract without the appropriate Triumph service tool 289.

3 The crankshaft pinion is also difficult to remove without Triumph service tool Z121. The chief danger is the risk of damage to the end of the crankshaft whilst the pinion is being drawn off; extreme care is necessary. If the Triumph service tool is not available, the pinion can be left in place until the crankcases are separated. It can then be removed by driving the crankshaft through the right hand side main bearing with a rawhide mallet. **This method is recommended only if the main bearing concerned is due for replacement.**

4 The idler pinion is easy to remove; it will pull off its centre spindle. Note how the timing pinions are marked; it will be necessary to re-align these marks in a certain manner during reassembly, to ensure the valve timing is correct. If it is desired to remove the camshaft pinions, mark the position of the key in relation to the keyway used before removing the pinions. This is IMPORTANT because each pinion has three keyways, only one of which is used.

5 Use Triumph service tool 289 to extract the pinions. They are a tight fit on the camshafts and there is no other way to remove them without risking damage. Screw on the extractor body until it is fully engaged with the threads of the pinion boss, then screw in the extractor bolt and turn clockwise. Extractor adaptor Z145 should be positioned within the end of the exhaust camshaft whilst the pinion is being withdrawn, to prevent damage to the taper in which the contact breaker cam seats.

10 Dismantling the engine - removing the primary drive

1 Remove the two domed nuts and copper washers from the forward end of the primary chaincase. Unscrew the eight crosshead screws around the periphery of the chaincase cover and withdraw the cover, complete with gasket.

2 Remove the three nuts retaining the stator coil assembly around the generator rotor and withdraw the assembly off the three studs. If the snap connectors of the electrical lead prevent the lead from pulling through the tunnel in the back half of the chaincase casting, unscrew the sleeve nut that threads into the chaincase. Removal of this nut will provide the extra clearance required.

3 To remove the rotor, bend back the tab washer and unscrew the retaining nut. The nut can be jarred off with a box or socket spanner and mallet, if the engine is locked by placing a metal bar through the eyes of the connecting rods. The rotor is keyed on to the crankshaft and should pull off easily.

4 As the primary chain is endless, it is necessary to withdraw the engine sprocket and clutch simultaneously. Slacken off the three adjuster nuts in the clutch pressure plate, using a penknife or some similar device to depress the clutch springs whilst the nuts are unscrewed. This is to free the springs from the three location pips on the underside of each nut which act as a self-locking device. A screwdriver, slotted as shown, will need to be used for the initial slackening.

5 Remove the adjusting nuts completely, then withdraw the pressure plate complete with the clutch springs and the cups in which they seat. The individual clutch plates, both plain and

inserted can now be drawn out, using two pieces of wire with a hook at the end.

6 When all the plates are withdrawn, access is available to the inner drum and shock absorber unit. With the engine still locked in position and a locking bar between the inner and outer clutch drums, unscrew the centre retaining nut and remove it, together with the cupped washer.

7 The complete clutch assembly can now be pulled off the clutch centre by inserting Triumph service tool D662/3 and screwing it home fully so that the full depth of thread engages. When the centre bolt is tightened, the clutch will be drawn off the mainshaft taper.

8 If the service tool is not available, the clutch pressure plate can be utilised to good effect. Refit it, without the clutch plates or the cups and springs, using the three adjuster nuts and washers placed beneath them, as shown in the accompanying photograph. (It will be necessary to file a notch in each washer, to accommodate the pip on the underside of each adjuster nut.) Unscrew the adjuster locknut in the centre of the clutch pressure plate and screw the adjuster inwards. It should pull the clutch assembly off the clutch centre splines when the end of the adjuster abuts on the end of the mainshaft.

9 The above technique should NOT be used if the clutch is an exceptionally tight fit on the clutch centre splines. Under these circumstances it is possible to distort the pressure plate permanently. Therefore the use of the correct Triumph service tool (D662/3) is advised.

10 Before the primary drive can be released, it is necessary to withdraw the engine sprocket. Triumph service tool Z151 is specified; the extractor bolts thread into the two tapped holes provided. If the tool is not available, there is room for the insertion of a sprocket puller provided the two forward-mounted stator coil retaining studs are removed first. When the taper of the engine sprocket is broken, the engine sprocket and clutch assembly can be drawn off their respective shafts, together with the primary chain, and then separated.

11 Note that if the Triumph service tool is not used, the clutch centre will remain on the gearbox mainshaft. The uncaged rollers of the clutch centre bearing will be displaced as the clutch chainwheel is withdrawn; they should be collected together and placed aside for reassembly.

12 The clutch centre can now be withdrawn, using a sprocket puller to break the keyed taper joint. If there is not sufficient access to engage the legs of the puller, mainshaft end float can be gained by removing the outer cover of the gearbox and releasing the nut retaining the kickstarter ratchet on the opposite end of the mainshaft. See Chapter 2, section 3.

11 Dismantling the engine - separating the crankcases

1 The crankcases can now be separated by removing the various studs and bolts that hold both crankcases together. Note that there are two screws within the crankcase mouth and a nut immediately to the rear of the timing case that must be removed. This means that it is necessary to remove both the outer and the inner gearbox covers, in order to gain access to this latter nut. See Chapter 2, sections 3 and 4 for the appropriate procedure. There is a bolt at the extreme end of the chaincase casting.

2 Pull the crankcase apart to release the crankshaft assembly. The left hand crankcase will have the inner portion of the primary chaincase attached containing the gearbox sprocket detachable cover. This need not be removed unless the cover is leaking or if the centre oil seal has to be renewed. It is retained by six countersunk screws which should be removed. The cover and gasket can then be pressed out of position, from the back of the chaincase.

3 The left hand crankcase will contain the outer race of the drive side roller bearing. This need not be disturbed, unless the bearing is to be renewed. Under these circumstances, the crankcase should be warmed to expand the bearing housing and release the outer race. It will drop free when the crankcase is brought down smartly on to a block of wood, outer face upwards. The oil seal that precedes the bearing will be left in position; it should be drifted out of location from the inside of the crankcase and renewed.

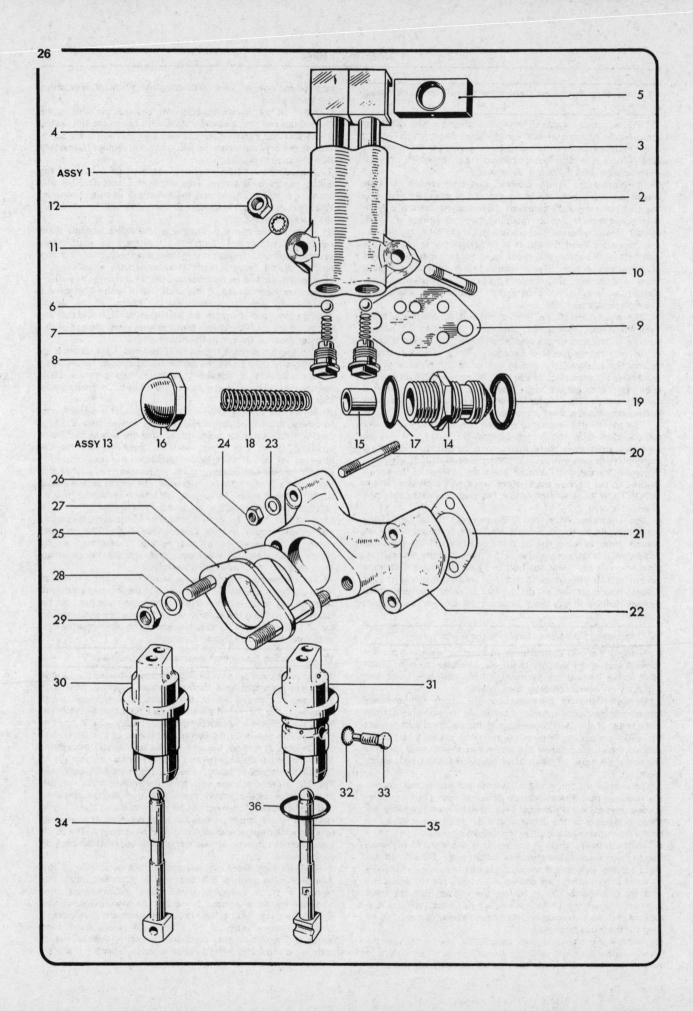

FIG.1.3 OIL PUMP, RELEASE VALVE, MANIFOLD (TR6) AND TAPPETS

1	Oil pump complete	19	Fibre washer
2	Pump body	20	Stud - 4 off
3	Feed plunger	21	Joint washer - 2 off
4	Scavenge plunger	22	Inlet manifold for TR6 model
5	Drive block	23	Plain washer - 4 off
6	Valve ball (7/16 in) - 2 off	24	Nut - 4 off
7	Valve spring - 2 off	25	Stud - 2 off
8	Screwed plug - 2 off	26	Joint gasket
9	Oil pump gasket	27	Heat insulator
10	Oil pump stud - 2 off	28	Plain washer - 2 off
11	Serrated washer - 2 off	29	Nut - 2 off
12	Nut - 2 off	30	Inlet tappet guide block
13	Pressure release valve complete	31	Exhaust tappet guide block
14	Valve body	32	Serrated washer - 2 off
15	Plunger	33	Set bolt - 2 off
16	Cap	34	Inlet tappet follower
17	Fibre washer	35	Exhaust tappet follower
18	Plunger spring	36	'O' ring seal

9.4 Make note of timing marks and keyways used, before dismantling

10.1 Eight screws and two nuts retain chaincase cover

10.2a Three nuts retain stator coil around crankshaft rotor

10.2b Unscrew sleeve nut if cable connectors will not pull through

10.3 Rotor is keyed fit on mainshaft

11.1a Inner and outer gearbox covers must be removed to gain access to rear crankcase nut

12 Examination and renovation - general

1 Now that the engine is stripped completely, clean all the component parts in a petrol/paraffin mix and examine them carefully for signs of wear or damage. The following sections will indicate what wear to expect and how to remove and renew the parts concerned, when renewal is necessary.

2 Examine all castings for cracks or other signs of damage. If a crack is found, and it is not possible to obtain a new component, specialist treatment will be necessary to effect a satisfactory repair.

3 Should any studs or internal threads require repair, now is the appropriate time. Care is necessary when withdrawing or replacing studs because the casting may not be too strong at certain points. Beware of overtightening; it is easy to shear a stud by overtightening giving rise to further problems, especially if the stud bottoms.

4 Where internal threads are stripped or badly worn, it is preferable to use a thread insert, rather than tap oversize. Most dealers can provide a thread reclaiming service by the use of Helicoil thread inserts. They enable the original component to be re-used.

11.1b Do not overlook bolt at end of chaincase casting

13 Main bearings and oil seals - examination and renovation

1 When the bearings have been pressed from their housings, wash them in a petrol/paraffin mix to remove all traces of oil. If there is any play in the ball or roller bearings, or if they do not revolve smoothly, new replacements should be fitted. The bearings should be a tight push fit on the crankshaft assembly and a press fit in the crankcase housings. A proprietary sealant such as Locktite can be used to secure the bearings if there is evidence of a slack fit and yet they are fit for further service.

2 The crankcase oil seal should be renewed as a matter of course, whenever the engine is stripped completely. This will ensure an oiltight engine.

11.1c ... or the two screws inside the crankcase mouth

14 Crankshaft assembly - examination and renovation

1 Wash the complete crankshaft assembly with a petrol/paraffin mix to remove all surplus oil. Mark each connecting rod and cap, to ensure they are replaced in exactly the same position, then remove the cap retainer nuts so that the caps and connecting rods can be withdrawn from the crankshaft. Keep the bolts and nuts together in pairs, so that they are replaced in their original order. It is best to unscrew the nuts a turn at a time, to obviate the risk of distortion.

2 Inspect the bearing surfaces for wear. Wear usually takes the form of scuffing or scoring, which will immediately be evident. Bearing shells are cheap to renew; it is wise to renew the shells if there is the slightest question of doubt about the originals.

3 More extensive wear will require specialist attention, either by having the crankshaft reground or by fitting a service-exchange replacement. If the crankshaft is reground, two undersizes of bearing shells can be obtained: —0.010 in and —0.020 in. The following table gives details of the various sizes of shell bearing and crankshaft, in relation to one another:

Shell bearing	Crankshaft size (limits)	
	in	mm
Standard	1.6235	41.237
	1.6240	41.250
Undersize:		
—0.010	1.6135	40.983
	1.6140	40.996
—0.020	1.6035	40.729
	1.6040	40.742

It is particularly important to note that the white metal bearing

11.2 When removed circular cover gives access to final drive sprocket

shells are prefinished to give the correct diametral clearance and on no account should the bearings be scraped or the connecting rod and cap joint filed in order to achieve a satisfactory fit. If such action seems necessary, the crankshaft has not been reground to the correct tolerances.

4   Check all connecting rod bolts for stretch by comparing the length with a new bolt. When the bolts are tightened fully to the recommended setting, stretch should not exceed 0.005 in.

5   It is not usually necessary to disturb the crankshaft assembly unless the lubrication system has become contaminated, in which case it may be advisable to clean out the central oil tube. Access is gained by unscrewing the retainer plug found in the right hand end of the crankshaft and then removing the flywheel bolt adjacent to the big end journal. The oil tube can be hooked out by passing a length of rod through the flywheel bolt orifice. Note that the retainer plug is retained by a centre punch mark and that it will be necessary to use an impact screwdriver, after the indentation has been drilled out.

6   Wash the oil tube in a petrol/paraffin mix and check that all the internal drillings in the crankshaft are quite free, before the tube is replaced. A jet of compressed air is best for this purpose. Make sure the retaining plug is tightened fully and centre punch the crankshaft at the screw slot, to retain the plug in position.

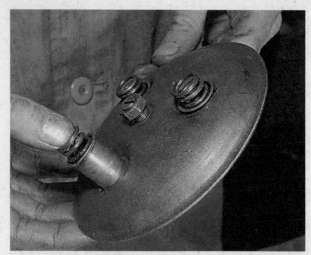

10.5a Remove clutch nuts to release pressure plate and springs

15 Camshaft and timing pinion bushes - examination and renovation

1   It is unlikely that the camshaft and timing pinion bushes will require attention unless the machine has covered a high mileage. The normal rate of wear is low. Bushes in the right hand crankcase can be removed by heating the crankcase to expand the surrounding metal and driving them out from the outside, using a two-diameter drift of the correct size. Fit the new bushes whilst the crankcase is still hot and make sure they are correctly aligned so that any oil feed holes register with those of the crankcase.

2   The blind bushes in the left hand crankcase are more difficult to remove. The recommended technique is to tap the bushes with a Whitworth thread and screw home a bolt of matching thread. If the crankcase is now heated, the bolt head can be gripped in a vice and the crankcase driven off the bush with a rawhide mallet. When the inlet camshaft bush is replaced, care must be taken to engage the peg with the breather porting disc that lies behind the bush. In the case of the exhaust camshaft bush, it will be necessary first to remove the tachometer drive assembly, as described in Section 22 (if fitted).

3   The camshaft bushes are machined from sintered bronze and only the smallest amount of metal will need to be removed after they are pressed into position. The correct internal bore diameters (fitted) are as follows:

10.5b Clutch plates can now be withdrawn

Left hand crankcase	0.8125 in - 0.8135 in
Right hand crankcase	0.874 in - 0.875 in

The intermediate (idler) timing gear bush is machined from phosphor bronze and should have an internal bore diameter (fitted) of 0.4990 in - 0.4995 in.

16 Camshafts, tappet followers and timing pinions - examination and renovation

1   Examine each camshaft, checking for wear on the cam form, which is usually evident on the opening flank and on the lobe. If the cams are grooved, or if there are scuff or score marks that cannot be removed by light dressing with an oilstone, the camshaft concerned should be renewed.

2   When extensive wear has necessitated the renewal of a camshaft, the camshaft and tappet followers should be renewed at the same time. It is false economy to use the existing camshaft followers with a new camshaft since they will promote a more rapid rate of wear.

10.6 With clutch and engine locked, remove centre nut

10.8 Clutch pressure plate used to extract clutch centre

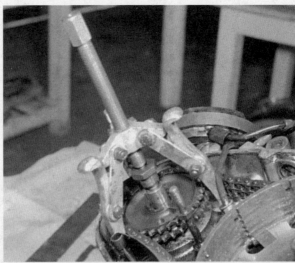

10.10 Sprocket puller used to detach sprocket together with chain and clutch

18.1 Press main bearings from crankcase for examination

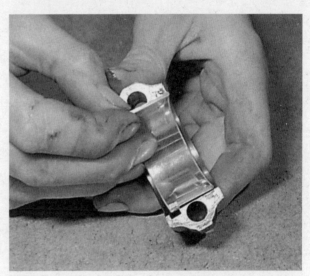

14.2a Replace bearing shells as a precaution

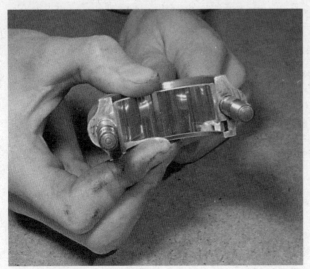

14.2b Shells locate with end caps as shown

14.3 Size of shells is clearly stamped on back

20.2a Collets are released when springs are compressed

20.2b Outer spring lifts off first ...

20.2c ... followed by inner spring

20.4 Use a double diameter drift to drive in new guides

20.6 Use a rotary action when grinding in the valves

25.1a Make sure shells are located correctly before tightening

3 Check each timing pinion for worn or broken teeth. Damage is most likely to occur if some engine component has failed during service and particles of metal have circulated with the lubrication system. Excessive backlash in the pinions will lead to noisy timing gear.

17 Cylinder barrel - examination and renovation

1 There will probably be a lip at the uppermost end of each cylinder bore that denotes the limit of travel of the top piston ring. The depth of the lip will give some indication of the amount of bore wear that has taken place, even though the amount of wear is not evenly distributed.
2 Remove the rings from the pistons, taking great care as they are brittle and easily broken. Most wear occurs within the top half of the bore, so the pistons should be inserted and the clearance between the skirt and the cylinder wall measured. If measurement by feeler gauge shows the clearance is 0.005 in greater or more than the figure quoted in the specifications section, the cylinder is due for a rebore. Oversize pistons are supplied in three sizes: + 0.010 in, + 0.020 in and + 0.040 in; the cylinders should be rebored to suit, as shown in the accompanying list of tolerances:

Piston size		Bore size			
		650 cc models		750 cc models	
in	mm	in	mm	in	mm
Standard		2.7948	70.993	2.9874	75.880
		2.7953	71.006	2.9882	75.9000
+ 0.010	† 0.254	2.8048	71.247	3.0021	76.2533
		2.8053	71.260	3.0010	76.2254
+ 0.020	+ 0.508	2.8148	71.501	3.0121	76.5073
		2.8153	71.514	3.0110	76.4794
+ 0.040	+ 1.016	2.8348	72.009	3.0321	76.9793
		2.8353	72.022	3.0310	76.9314

3 Give the cylinder barrel a close visual inspection. If the surface of either of the bores is scored or grooved, indicative of a previous engine seizure or a displaced circlip and gudgeon pin, a rebore is essential. Compression loss will have a very marked effect on performance.
4 Check that the outside of the cylinder barrel is clean and free from road dirt. Use a wire brush on the cooling fins if they are obstructed in any way. The application of matt cylinder black will help improve the heat radiation.
5 Check that the base flange is not cracked or damaged. If the engine has been overstressed, one of the first parts to fail is the base of the cylinder barrel, either at the holding down points or around the base of each bore. If a crack is found, the cylinder barrel should be renewed.
6 The rebore limit is +0.040 in. Above this size, the cylinder walls cannot be considered to have sufficient thickness consistent with safety and reliability. Re-sleeving or a service exchange replacement is the only practicable solution to the problem.

18 Pistons and piston rings - examination and renovation

1 Attention to the pistons and piston rings can be overlooked if a rebore is necessary, since new replacements will be fitted.
2 If a rebore is not considered necessary, examine each piston closely. Reject pistons that are scored or badly discoloured as the result of exhaust gases by-passing the rings.
3 Remove all carbon from the piston crowns, using a blunt scraper which will not damage the surface of the piston. Clean away all carbon deposits from the valve cutaways and finish off with metal polish so that a clean, shining surface is achieved. Carbon will not adhere so readily to a polished surface.
4 Check that the gudgeon pin bosses are not worn or the

circlip grooves damaged. Check that the piston ring grooves are not enlarged. Side float should not exceed 0.003 in.
5 Piston ring wear can be measured by inserting the rings in the bore from the top and pushing them down with the base of the piston so that they are square in the bore and about 1½ inches down. If the end gap exceeds 0.014 in (all rings) renewal is necessary.
6 Check that there is no build up of carbon on the inside surface of the rings or in the grooves of the pistons. Any build-up should be removed by careful scraping.
7 The piston crowns will show whether the engine has been rebored on some previous occasion. All oversize pistons have the rebore size stamped on the crown. This information is essential when ordering replacement piston rings.

19 Small end bearings - examination and renovation

1 The amount of wear in the small end bushes can be ascertained by the fit of the gudgeon pins. The pin should be a good sliding fit in each case, without evidence of any play.
2 Renewal can be effected by using a simple drawbolt arrangement (as illustrated) whereby the new bush is used to press the old bush out of location. It is essential to ensure the oilway in the bush locates with the oilway in the connecting rod, otherwise the bearing will run dry and rapid wear will occur.
3 After the bushes have been fitted, they will have to be reamed out to the correct size. Cover the mouth of the crankcase with rag to prevent metallic particles from dropping in, and ream out to 0.6890 to 0.6894 in.

20 Cylinder head and valves - dismantling, examination and renovation

1 It is best to remove all carbon deposits from the combustion chambers, before removing the valves for grinding-in. Use a blunt-ended scraper so that the surface of the combustion chambers is not damaged and finish off with metal polish to achieve a smooth, shiny surface.
2 Before the valves can be removed, it is necessary to obtain a valve spring compressor of the correct size. This is necessary to compress each set of valve springs in turn, so that the split collets can be removed from the valve cap and the valve and valve spring assembly released. Keep each set of parts separate; there is no fear of inadvertently interchanging the valves because the heads are marked 'IN' or 'EX'.
3 Before giving the valves and valve seats further attention, check the clearance between each valve stem and the valve guide in which it operates. Some play is essential in view of the high temperatures involved, but if the play appears excessive, the valve guides must be renewed.
4 To remove the old valve guides, heat the cylinder head and drive them out of position with a double diameter drift of the correct size. Replace the new guides, whilst the cylinder head is still warm. Note that when bronze valve guides are fitted, the two SHORT guides are fitted in the inlet position. Conversely, when cast iron guides are fitted, the two LONG guides are fitted in the inlet position.
5 Grinding in will be necessary, irrespective of whether new valve guides have been fitted. This action is necessary to remove the indentations in the valve seats caused under normal running conditions by the high temperatures within the combustion chambers. It is also necessary when new valve guides have been fitted, in order to re-align the face of each valve with its seating.
6 Valve grinding is a simple task. Commence by smearing a trace of fine valve grinding compound (carborundum paste) on the valve seat and apply a suction tool to the head of the valve. Oil the valve stem and insert the valve in the guide so that the two surfaces to be ground in make contact with one another. With a semi-rotary motion, grind in the valve head to the seat, using a backward and forward action. Lift the valve occasionally so that the grinding compound is distributed evenly. Repeat the

operation until a ring of light grey matt finish is obtained on both valve and seat. This denotes the grinding operation is complete. Before passing to the next valve, make sure that all traces of compound have been removed from both the valve and its seat and that none has entered the valve guide. If this precaution is not observed, rapid wear will take place due to the abrasive nature of the carborundum base.

7 When deeper pit marks are encountered, or if the fitting of a new valve guide makes it difficult to obtain a satisfactory seating, it will be necessary to use a valve seat cutter set to an angle of 45° and a valve refacing machine. This course of action should be resorted to, only in an extreme case, because there is risk of pocketing the valve and reducing performance. If the valve itself is badly pitted, fit a replacement.

8 Before reassembling the cylinder head, make sure that the split collets and the taper with which they locate on each valve are in good condition. If the collets work loose whilst the engine is running, a valve will drop and cause extensive engine damage. Check the free length of the valve springs with the specifications section and renew any that have taken a permanent set.

9 Reassemble by reversing the procedure used for dismantling the valve gear. Do not neglect to oil each valve stem before the valve is replaced in the guide.

10 Before setting aside the cylinder head for reassembly, make sure that the cooling fins are clean and free from road dirt. Check that no cracks are evident, especially in the vicinity of the holes through which the holding down studs and bolts pass, and near the spark plug threads.

11 Finally, make sure that the cylinder head flange is completely free from distortion at the joint it makes with the cylinder barrel. An aluminium alloy cylinder head will distort with comparative ease if it is tightened unevenly and may lead to a spate of blowing cylinder head gaskets. If the amount of distortion is not too great, flatness can be restored by careful rubbing down on a sheet of fine emery cloth wrapped around a sheet of plate glass. Otherwise it may be necessary to have the cylinder head flange refaced by a machining operation.

21 Tappets and tappet guide blocks - examination and renovation

1 Mention has not been made of the tappets or tappet guide blocks, which seldom require attention. The amount of wear within the tappet blocks can be ascertained by rocking the tappet whilst it is within the tappet block. It should be a good sliding fit, with very little sideways movement.

2 To remove and replace the tappet blocks, first remove the locking screw. The block can then be drifted out of position, preferably by using Triumph service tool Z23. An O ring seal is inserted between each tappet block and the cylinder barrel, which should be renewed as a matter of course.

3 To replace the tappet blocks, first grease the outer surface, then fit the O ring and align the hole for the locking screw before driving the block back into position, with the shoulder flush against the flange. Triumph service tool Z23 is again recommended to simplify this operation.

4 As mentioned in Section 16.2, the tappet followers should be renewed if they show signs of wear. Make sure that the replacements are fitted correctly, with the machined cut-away in the stem facing the outside of the tappet guide block. If fitted incorrectly, the tappets will not be lubricated. It is also important to note that the inlet and exhaust tappets and the tappet guide blocks must not be interchanged.

5 From engine no DU 24875, oil has been fed to the rubbing surfaces of the cam followers and it is important that the correct tappet followers are fitted. Several changes in the lubrication system make it essential to identify and fit the correct parts. If conversion to the latest exhaust tappet followers is required on machines from engine nos DU 24876 to DU 63043, it is essential that metering jet E6348 is removed from the dowel in the front timing cover and replaced with hollow dowel T989. Exhaust tappet follower E6329 should be replaced by part

number E8895 in the case of 6T and TR6 models and exhaust tappet follower E6490 by part number E8801, in the case of the T120 models.

22 Removing and replacing the tachometer drive

1 On models fitted with a tachometer, the drive is taken from the exhaust camshaft left hand end. The cable is connected directly to a crankcase-mounted adaptor on 1963 to 1965 models, the drive being transmitted by a spade from the camshaft. On 1966 and later models, the cable is connected to a separate gearbox which turns the drive through 90° and gives a 2:1 reduction ratio. Note that the later assembly can be fitted to earlier models.

2 To remove the gearbox, disconnect the cable and unscrew the large slotted end cap. Extract the drive pinion, either using long-nosed pliers or by turning the engine over quickly. This will reveal the central retaining sleeve bolt, which must be unscrewed using a slim box spanner or deep socket. On models with engine nos. DU24875 to DU85903 (1966 to 1968) a $\frac{3}{8}$ in Whitworth spanner will be required and the bolt has a normal right hand thread. Later models from engine no DU85904 (1969 on) will require a $\frac{7}{16}$ in AF (approx $\frac{3}{16}$ in Whitworth) spanner and the bolt has a left hand thread, ie it is unscrewed clockwise. Withdraw the gearbox, noting the special sealing washer (fitted as standard to all models from engine no. DU44394); this is fitted to prevent slackening and oil leakage, and must be renewed whenever the gearbox is disturbed.

3 To remove the driven gear, first unscrew the locating set screw (early models) or drive out the locking pin (later models) and withdraw the driven gear housing, noting that it will be a tight fit; note the sealing O-ring. Remove the driven gear from its housing. Ensure that all O-rings and seals are renewed on reassembly.

4 On models up to 1970 (engine no. JD24996) a slotted thimble is pressed (approx $\frac{1}{16}$ in) into the exhaust camshaft left hand end to transmit drive to the spade. If the thimble is damaged it can be renewed from outside the crankcases, providing care is taken to extract as much of the old part as possible and to locate the new part carefully so that the drive spade can engage it fully. On later models a slotted plug is permanently fixed in the camshaft end.

23 Pushrods, rocker spindles and rocker arms - examination and renovation

1 Check the pushrods for straightness by rolling them on a sheet of glass. If any are bent, they should be renewed since it is not easy to effect a satisfactory repair.

2 Check the end pieces to ensure that they are a tight fit on the light alloy tubes. If the end pieces work loose, the pushrod must be renewed. It is unlikely that the end pieces will show signs of wear at the point where they make contact with the rocker arms and the tappet followers, unless the machine has covered a very high mileage. Wear usually takes the form of chipping or breaking through the hardening, which will necessitate renewal.

3 Examine the tips of the rocker arms. If wear is evident, both the valve clearance adjusters and the ball pins should be renewed. The latter should be pressed into place with the drilled flat towards the rocker spindle.

4 If it is necessary to renew the rocker spindles, they can be driven out of the rocker box housing by means of a soft metal drift. Reassembly is best accomplished by using either Triumph service tool Z111 or a 7/16 in bolt six inches long with a taper ground on the end. The accompanying illustration shows the way in which the various washers are assembled. Note that each spindle has a plain washer with a small diameter bore that acts as a thrust washer and is assembled last against the inner right hand face of the rocker box. The oil feed to the rocker spindles must be on the right hand side of the machine. Before fitting the spindles, check that the oilways are clean and unobstructed,

preferably by using a jet of compressed air. Lubricate the spindle thoroughly before it is inserted and make sure it has a new O ring seal.

5 There is no adjustment for end float. This function is performed by the spring washers, fitted between each end of the rocker arm and the rocker box cover.

24 Engine reassembly - general

1 Before the engine is reassembled, all the various components must be cleaned thoroughly so that all traces of old oil, sludge, dirt and gaskets etc are removed. Wipe each part with clean, dry lint-free rag to make sure there is nothing to block the internal oilways of the engine during reassembly.

2 Make sure that all traces of the old gaskets have been removed and that the mating surfaces are clean and undamaged. One of the best ways to remove old gasket cement is to apply a rag soaked in methylated spirit. This acts as a solvent and will ensure the cement is removed without resort to scraping and subsequent risk of damage. Special care should be taken with regard to the crankcases to ensure that the locating dowels are positioned correctly. The metering dowel fitted in the timing cover case, close to the pressure release valve, incorporates a metering pin to control the flow of oil to the exhaust tappet followers and must be assembled with the larger bore facing OUTWARDS.

3 Gather together all the necessary tools and have available an oil can filled with clean engine oil. Make sure the new gaskets and oil seals are to hand; nothing is more infuriating than having to stop in the middle of a reassembly sequence because a vital gasket or replacement has been overlooked.

4 Make sure that the reassembly area is clean and that there is adequate working space. Refer to the torque and clearance settings whenever they are given. Many of the smaller bolts are easily sheared if they are overtightened. Always use the correct size screwdriver bit for the crosshead screws and never an ordinary screwdriver or punch.

25 Engine reassembly - rebuilding the crankshaft

1 Refit the connecting rods to the crankshaft assembly in their original positions, using the marks made during the dismantling operation as a guide. Make sure that the shell bearings are located correctly, then replace the end caps, the retaining nuts and bolts. Tighten each nut evenly until the connecting rods and their end caps seat correctly, then tighten the nuts with a torque spanner to a load of 28 ft lb. Check that both connecting rods revolve freely, with an absolute freedom from play.

2 Apply a pressure oil can to the drilling at the right hand end of the crankshaft and pump until oil is expelled from both big ends. This is essential, to ensure that the oil passages are free from obstruction and full of oil.

3 If a torque spanner is not available, tighten the connecting rod nuts until a micrometer reading shows the bolts have stretched to a maximum of 0.005 in. This extension figure is to be preferred as a means of accurately tensioning the bolts without risk of overstress.

26 Engine reassembly - reassembling the crankcases

1 If the main bearings have been removed, replace the ball journal bearing in the right hand crankcase and the outer race of the roller bearing in the left hand crankcase. It is advisable to heat the respective crankcases beforehand so that the bearings will drop into place without difficulty.

2 The inner race of the roller bearing should be driven on to the left hand side of the crankshaft assembly until it is hard against the shoulder of the crankshaft.

3 Any shims that have been fitted to limit the end float of the crankshaft assembly must be fitted BEHIND the main bearings in the crankcases. They should be of equal thickness on either side, so that the crankshaft assembly is centrally disposed within the crankcase.

4 Mount the left hand crankcase on two blocks of wood so that there will be sufficient clearance for the end of the crankshaft to project downwards without touching the bench. Lubricate the main bearings and the camshaft bushes and place the rotary breather valve and spring in the inlet camshaft bush. If the camshafts are not fitted to the right hand crankcase, position them in their respective bushes taking care that the slot in the end of the inlet camshaft engages with the projection of the rotary breather valve and the slot of the exhaust camshaft with the spade of the tachometer drive (if fitted).

5 Lower the crankshaft assembly into position and give it a sharp tap to ensure that the inner race of the roller bearing is fully engaged with the outer race.

26.4 Do not omit to insert rotary breather and spring in left hand camshaft bush

26.5 Lower crankshaft assembly into left hand crankcase first

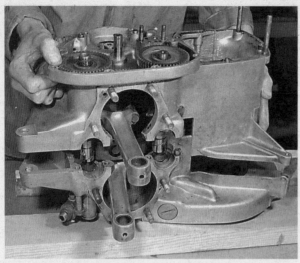

26.6 Lower right hand crankcase and register components with bearings

27.1a Washer must precede fitting of crankshaft pinion

27.1b. Do not omit key before driving pinion into position

27.5 Check alignment of holes when fitting new oil pump gasket

27.6a Fit new crankshaft oil seal and ...

27.6b ... circlip that holds seal in position

27.6c Contact breaker assembly also has oil seal

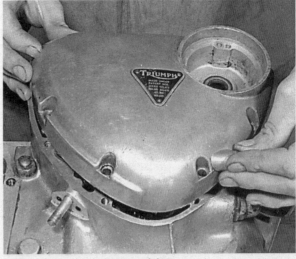

27.7 No gasket at timing cover joint

28.2 Pipe on right is main feed pipe

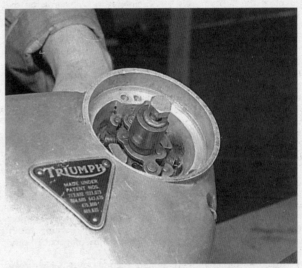

28.5. Auto-advance assembly is keyed on to exhaust camshaft

29.3 Warm pistons to make insertion of gudgeon pins easier

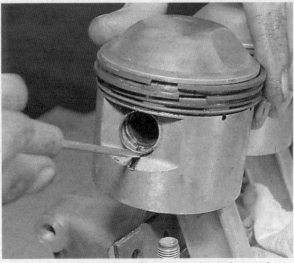

29.4 Always double-check that circlips are located correctly

6 Coat the jointing face of the right hand crankcase with gasket cement and lower it into position, after checking that the connecting rods are centrally disposed. If the camshafts are still attached to the right hand crankcase it will be necessary to rotate them until they engage with the rotary breather valve and the tachometer drive (if fitted) before the crankcases will meet. Push both crankcases together so that they mate correctly all round and check that the crankshaft and the camshafts will revolve quite freely before the securing bolts and studs are replaced and tightened by hand.

7 Check that the cylinder barrel junction of the crankcases is level and if necessary adjust by light tapping. When a level surface is achieved and the crankshaft and camshafts are free to revolve, the securing bolts can be tightened fully with a torque spanner to 20 ft lb (studs) and 13 ft lb (bolts). The accompanying illustration shows the bolts and screws that should be tightened first. (See Fig. 1.8, page 47)

27 Engine reassembly - replacing the timing pinions, oil pump and timing cover

1 Place the clamping washer, chamfered side inwards, on the end of the crankshaft. Replace the Woodruff key in the right hand end of the crankshaft and slide the crankshaft pinion into position so that the keyway locates with the key. Drive the pinion on to the end of the crankshaft until it is fully home.

2 If the camshaft pinions have been removed, position the keys in the respective camshafts which will have been marked to ensure that the correct keyways are used. Replace the pinions, again ensuring that the previously marked keyway is the one aligned. Triumph service tool Z89 can be used for replacing the pinions, in conjunction with replacer adaptor Z144. Alternatively, the pinions can be drifted on to their respective shafts using a hollow tube. Replace the pinion locknuts, noting that each has a left hand thread. The inlet camshaft nut has the projecting peg for the oil pump drive.

3 Replace the idler pinion on its shaft and align the pinions so that the timing marks correspond EXACTLY as shown in the accompanying diagram. If these marks align correctly, the valve timing is correct. It is best to remove and replace the idler pinion when aligning the timing marks because the hunting tooth principle used means the marks will coincide only once every 94th revolution.

4 Replace the crankshaft pinion securing nut which may otherwise obscure the timing mark of the pinion. This has a normal right hand thread and should be tightened fully.

5 The oil pump can now be replaced as a complete unit. Fit a new gasket between the oil pump body and the oilways in the timing case, making sure that the holes in the gasket align correctly. DO NOT USE GASKET CEMENT AT THIS JOINT. Thread the pump body over the two retaining studs and engage the driving peg of the inlet camshaft pinion retaining nut with the hole in the sliding block at the top of the pump. Make sure the pump body lies flat against the crankcase without any strain and replace the two conical retaining nuts. The conical portion of the nuts should face inwards; tighten the nuts with a torque wrench to 5 ft lbs.

6 Before replacing the timing cover, renew the oil seals surrounding the extreme end of the crankshaft and the end of the exhaust camshaft which drives the contact breaker cam. The former is retained by a circlip which must be replaced correctly after the new oil seal has been pressed into position.

7 The timing cover has no gasket - a face to face joint is used. Make sure both mating surfaces are perfectly clean, then apply a thin smear of gasket cement to ensure a satisfactory, oiltight seal. Check that both dowels are in position and correctly located (dowel containing the metering valve nearest to the pressure release valve, larger bore facing OUTWARDS) and fit the timing cover, making sure it engages correctly with the end of the crankshaft. Insert the eight crosshead screws securing the cover in position and tighten them fully. The three long

screws are located at the extreme left hand corner of the cover and in the lower left and right hand positions.

28 Engine reassembly - replacing the oil feed pipes and the gearbox end covers

1 It is advisable to replace the oil feed pipes attached to the underside of the timing cover at this stage because they are difficult to fit after the gearbox end covers are in position. Fit a new gasket at the union joint (NO gasket cement) and slide the union over the projecting stud. Replace the retaining nut and washer and tighten down.

2 Confusion often arises about the identification of the pipes, neither of which is marked. The forward-mounted of the two pipes carries the main oil feed from the oil tank. The rear-mounted pipe is the scavange or return pipe.

3 Refit the gearbox inner cover following the procedure described in Chapter 2, section 14. It is essential that the gear selector mechanism is 'timed' in the manner described when the end cover is positioned, otherwise it will be impossible to select the gears correctly.

4 Fit the gearbox outer cover, following the procedure given in Chapter 2, section 15.

5 Replace the contact breaker auto-advance assembly and contact breaker cam, which is keyed to the end of the exhaust camshaft to aid correct location. Insert the centre retaining bolt and tighten it fully.

6 As a temporary expedient to obviate the risk of damage whilst the engine reassembly proceeds, fit the end cover of the contact breaker assembly, which is retained by two crosshead screws.

29 Engine reassembly - replacing the pistons and cylinder barrel

1 Stand the engine upright and arrange the crankshaft so that both connecting rods are fully extended. Add about 1/6 pint of engine oil to the crankcase, pad the mouth of the crankcase with clean rag to prevent any displaced parts from falling in; commence engine reassembly.

2 Oil both gudgeon pins and small end bushes. Check that the oilway in each small end bush lines up with the oil hole in the connecting rod.

3 Warm both pistons in hot water to make the insertion of the gudgeon pin easier. Make sure the pistons are replaced in their original positions, using the markings made previously within the skirt. Do not reverse them, back to front.

4 After inserting the gudgeon pins and circlips, check that each circlip is located positively in its groove. A displaced circlip will cause severe engine damage if it works loose whilst the engine is running.

5 Always fit new circlips. Never re-use the originals.

6 Fit a new cylinder base gasket (NO gasket cement) and check to ensure that the oil holes line up with the drillings in the crankcase flange. Failure to observe this precaution will result in the tappets being starved. Check that both dowels in the left hand crankcase are positioned correctly.

7 Replace the piston rings. The compression rings are tapered and it is important to ensure the land marked 'top' is fitted uppermost in each case. When the rings are fitted, space them out so that the end gaps do not coincide and fit a pair of piston ring clamps. Place two pieces of wood below the skirt of each piston and rotate the crankshaft so that the pistons seat on these 'stops' to hold them steady.

8 Oil the tappet followers in the cylinder barrel and retain them to the cylinder barrel with a rubber band or by forcing a rubber block between them. Oil the cylinder bores and lower the cylinder barrel on to the pistons so that each engages with its respective bore and the piston ring clamps are displaced.

9 When the piston rings have engaged fully with the bores, remove the piston ring clamps and the rag used to pad the crankcase. Remove the 'steadies' below each piston and lower the cylinder barrel until it seats over the retaining studs. Replace

29.6 Fit new cylinder base gasket

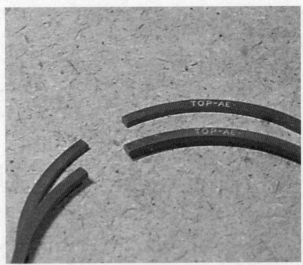

29.7a Piston rings are clearly marked

29.7b Note use of rods as piston steadies

29.8a Secure tappets with rubber bands before fitting barrel

29.8b Oil cylinder bores and pistons prior to assembly

29.8c Piston ring clamps make assembly much easier

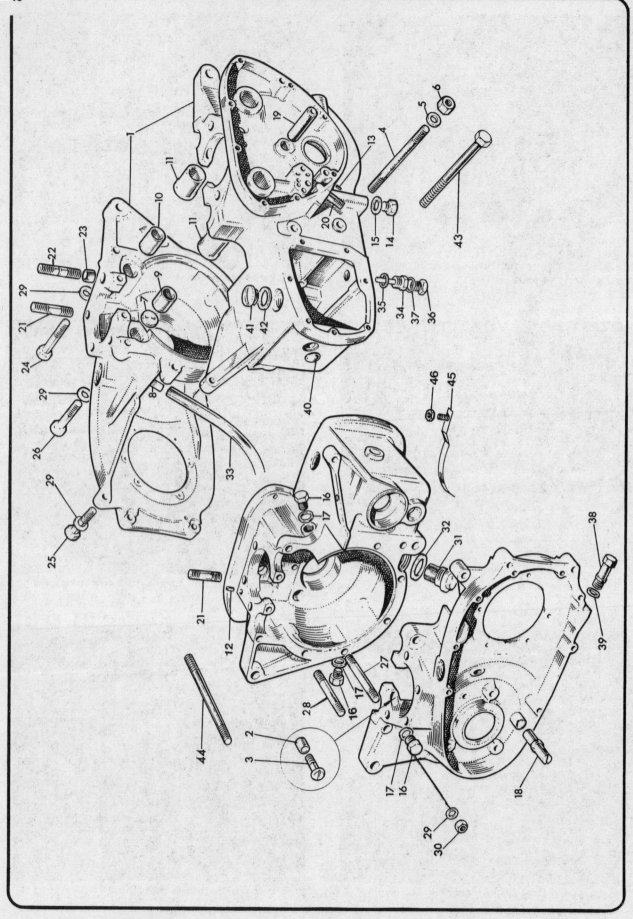

FIG.1.4 CRANKCASE COMPONENTS

1 Crankcase assembly complete

2 Dowel (hollow)

3 Screw - 2 off

4 Stud

5 Plain washer - 2 off

6 Nut - 2 off

7 Breather disc

8 Breather pipe

9 Inlet camshaft bush

10 Exhaust camshaft bush

11 Camshaft bush, inlet and exhaust - 2 off

12 Hollow dowel, tappet oil feed

13 Locating peg

14 Oilway plug

15 Copper washer

16 Tachometer drive plug - 2 off

17 Fibre washer - 2 off

18 Chain tensioner abutment

19 Idler pinion spindle

20 Oil pipe stud

21 Cylinder stud - 6 off

22 Cylinder stud - 2 off

23 Hollow dowel - 2 off

24 Bolt

25 Bolt

26 Bolt

27 Stud

28 Stud - 2 off

29 Plain washer - 11 off

30 Nut - 8 off

31 Crankcase filter

32 Copper washer

33 Breather extension pipe

34 Drain plug with level tube

35 Fibre washer

36 Level plug

37 Copper washer

38 Drain and adjuster plug

39 Fibre washer

40 Blanking plug

41 Filler plug - 2 off

42 Fibre washer - 2 off

43 Bolt

44 Stud

45 Protection blade

46 Self-locking nut

42

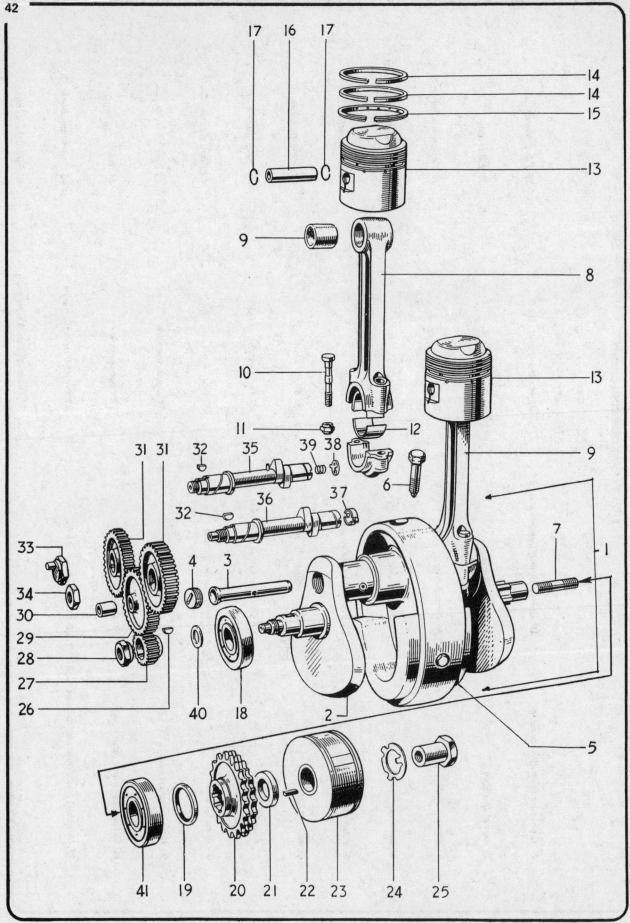

FIG. 1.5. CRANKSHAFT AND CONNECTING RODS

1 Crankshaft and flywheel assembly complete	11 Self locking nut - 4 off	21 Distance piece	32 Camshaft key - 2 off
2 Crankshaft complete with oil tube	12 Big end bearing shell - 4 off	22 Rotor key	33 Inlet camshaft nut
3 Oil tube	13 Piston complete - 2 off	23 Rotor, Lucas RM19	34 Exhaust camshaft nut
4 Screwed plug	14 Taper compression ring - 4 off	24 Tab washer	35 Inlet camshaft
5 Flywheel	15 Oil control ring - 2 off	25 Nut	36 Exhaust camshaft
6 Flywheel bolt - 3 off	16 Gudgeon pin - 2 off	26 Key	37 Tachometer drive thimble
7 Crankshaft stud	17 Circlip - 4 off	27 Timing pinion (crankshaft)	38 Rotary breather valve
8 Connecting rod - 2 off	18 Main bearing, right	28 Nut	39 Breather valve spring
9 Small end bush - 2 off	19 Oil seal	29 Idler pinion	40 Clamping washer
10 Connecting rod bolt - 4 off	20 Engine sprocket, 29 teeth duplex	30 Idler pinion bush	41 Main bearing, left
		31 Camshaft pinion - 2 off	

29.11 Check engine revolves freely by holding barrel and turning crankshaft

30.1a New pushrod tube oil seals are essential

30.1b Fit new oil seals at upper ends too

30.3a Old cylinder head gasket can be re-used if annealed

30.3b Lower cylinder head and fit bolts, finger-tight only

31.3 Contact breaker cover must have gasket to seal from damp

and tighten the cylinder base nuts.

10 If piston ring clamps are not available, it is possible to compress and feed in the piston rings by hand, although this is an operation that requires two persons - one to hold the cylinder barrel and the other to feed in the pistons and rings. A chamfer at the bottom of each cylinder bore aids the engagement of the piston rings.

11 Check that the engine revolves quite freely before proceeding further.

30 Engine reassembly - replacing the pushrod tubes and cylinder head

1 Position the two pushrod tubes at the front and rear of the cylinder barrel, after removing the retainers used to hold the tappet followers in position. Each tube should have a new oil seal fitted to each end.

2 Reassemble the cylinder head by reversing the dismantling procedure recommended for removing the valves. Check that the split collets are located correctly within each valve spring cap - a light tap with a hammer on the end of each valve stem is a good check.

3 Place a new copper cylinder head gasket on top of the cylinder barrel and lower the cylinder head into position, making sure the pushrod tubes locate with their respective tunnels. Check that the tubes are aligned correctly, as shown in the accompanying illustration, otherwise there is risk of the pushrods fouling. When the bolt holes are aligned correctly with the cylinder head gasket, fit the outer holding-down bolts and the bolt in the centre of the cylinder head (two Allen screws, 750 cc models). They should be fingertight only at this stage of assembly.

31 Replacing the contact breaker assembly and timing the ignition

1 Remove the contact breaker cover which was temporarily replaced and insert the automatic advance mechanism and the contact breaker cam. The former engages with an extension of the exhaust camshaft and is secured by the centre bolt which also retains the cam. Replace the contact breaker assembly secured by two pillar bolts.

2 It is convenient, whilst the engine is still on the bench, to check and if necessary reset the ignition timing. Different procedures are required according to whether or not a stroboscope is available and whether the engine number is prior to, or later than, DU 66246. Refer to Chapter 5, section 7 for the relevant detailed information.

3 The static method of ignition is preferred amongst owners even though it is less accurate. Comparatively few have access to a stroboscope or instruction in its use. Provision is made in the engine design to make timing by the static method as simple, yet as accurate, as possible.

32 Reassembling the engine - replacing the gearbox final drive sprocket

1 If the gearbox has not been included in the overhaul, the final drive sprocket will have remained in situ, since its presence in no way hinders the dismantling of the engine. If, however, the sprocket has been withdrawn for any reason, it must be replaced before the gearbox mainshaft is inserted and the primary transmission is built up.

2 The sprocket fits on the splined sleeve gear pinion of the mainshaft and is locked in position by a large diameter nut (right hand thread) and tab washer. On current models, it is customary to apply a compound known as Hydroseal to the sleeve gear splines before fitting the sprocket, to act as additional locking. It is essential that the sprocket is locked up tight; if it works loose, the splines on the sleeve gear pinion will wear away very rapidly.

3 The rear cover plate can now be attached to the chaincase casing integral with the left hand crankcase. Before the

33.2 Engine sprocket, clutch and chain must be replaced simultaneously

33.5 Replace two forward-mounted studs when sprocket is in position

33.6 Use torque wrench for tightening crankshaft nut

cover is fitted, the oil seal in the centre should be replaced. This is a vital seal because it prevents the escape of oil along the gearbox mainshaft when the engine is running and the oil content of the chaincase is widely distributed. There is a paper gasket between the cover plate and the chaincase; the jointing faces should be smeared with gasket cement to ensure a good, oiltight seal, before the retaining screws are replaced and tightened.

33 Engine reassembly - replacing the clutch, engine sprocket and primary chain

1 The primary chain fitted to the unit construction twins is of the duplex type. No spring link is fitted; the chain must be fitted together with the engine sprocket and clutch, at one and the same time. Before commencing assembly, slacken off the chain tensioner by turning in an anticlockwise direction the adjuster within the tunnel in the bottom of the chaincase.

2 Replace the clutch centre complete with roller bearings and clutch chainwheel after the Woodruff key has been inserted in the gearbox mainshaft, together with the engine sprocket and chain loop. The engine sprocket must be positioned so that the taper ground boss is closest to the crankshaft main bearing; the sprocket is a light drive fit on the crankshaft splines. Tap the clutch centre to lock it on to the gearbox mainshaft.

3 Fit the clutch inner drum over the splines of the clutch centre and replace the cup washer and self-locking nut that retains the assembly in position. Scotch the clutch inner drum and tighten the centre nut with a torque wrench to a setting of 50 ft lb. Machines having an engine number prior to DU 48144 have a slightly different locking arrangement embodying a tab washer for locking the centre nut after it has been tightened.

4 Replace the clutch plates, commencing with an inserted plate and then following with alternate plain and inserted plates. Fit the domed pressure plate, after the clutch pushrod has been inserted in the centre of the hollow mainshaft, and replace the

thimbles, clutch springs and clutch adjusting nuts. Tighten the adjusting nuts evenly until at least one half of the thread has engaged with the stud projecting from the clutch inner drum.

5 Replace the two forward-mounted studs which were removed to make the withdrawal of the engine sprocket easier and the sleeve nut in the chaincase rear, through which the leads from the generator stator coils pass. Fit the distance piece, located between the engine sprocket and the generator rotor, then position the key in the left hand end of the crankshaft and slide the generator rotor into position after aligning the keyway. Refit the tab washer and the centre retaining nut and tighten the latter with a torque wrench to a setting of 30 ft lb. Bend the tab washer to lock the nut.

6 Slip the distance pieces over the stator coil retaining studs and position the stator coil assembly so that the lead connecting the coils is at the top. Thread the lead through the sleeve nut in the rear of the chaincase, until it emerges from the back. Replace the split plug that seals the sleeve nut orifice and the rubber cap over the sleeve nut extension. Check that there is no possibility of the lead fouling the primary chain. The lead must emerge from the outer portion of the stator coil assembly and be held in the clip by the rear of the engine sprocket.

7 Replace the stator plate retaining nuts and washers, then check that the rotor does not foul the stator plate assembly when the engine is turned over. There should be a minimum clearance of 0.008 in between each of the stator coil pole pieces and the rotor.

8 Refit the primary chaincase outer cover temporarily to

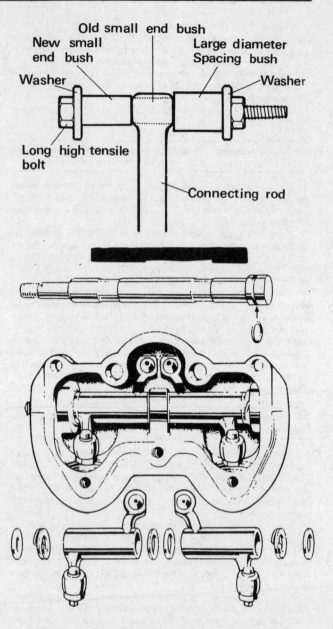

Fig.1.7. Rocker box assembly

prevent damage whilst the engine unit is lifted back into the frame. Replace the spark plugs to prevent dirt or other foreign matter from entering the engine during this same operation.

34 Replacing the engine unit in the frame

1 Replacement of the engine unit is a two man operation, unless a garage hoist is available. The unit should be fitted from the right hand side of the machine and lowered into approximately the correct position.

2 Replace the front engine bolt first, to secure the engine unit in approximately the correct position. Then insert the bolt below the engine, which has a distance piece each side of the crankcase lug.

3 Fit the rear engine plates and the four nuts and bolts that secure each to the engine unit and frame. Replace the centre nut and washer on the engine plate location stud.

4 Tighten the engine bolts fully. If they work loose or are not tightened sufficiently excess engine vibration will be experienced.

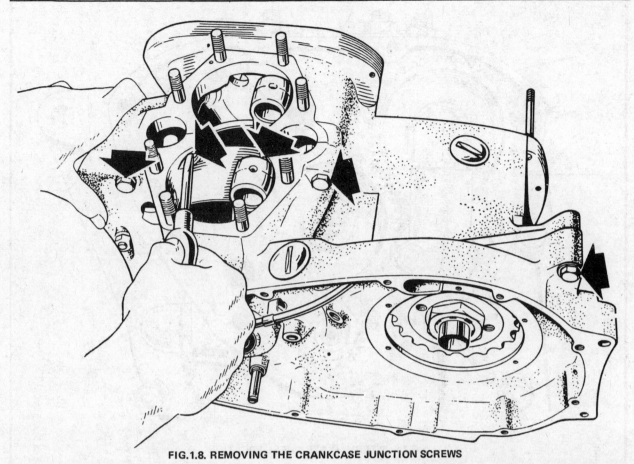

FIG.1.8. REMOVING THE CRANKCASE JUNCTION SCREWS

35 Engine reassembly - replacing the pushrods and rocker boxes

1 Now that the engine has been installed in the frame, the pushrods and rocker boxes can be added so that the engine rebuild is complete. Insert the pushrods in their original positions by following the markings made at the time when the engine was dismantled. They must engage correctly with the ends of the tappet followers, one of which will be lifted. Fit new rocker box gaskets, using a thin smear of gasket cement and after rotating the engine so that the inlet pushrods are at the bottom of their stroke, fit and tighten down the inlet rocker box so that the bolts are fingertight only. Do not forget the three nuts on the underside of the casting.

2 Before tightening down, it is important to check that each pushrod end has located correctly with the ball end of its respective rocker arm. When the inlet rocker box is tightened down, rotate the engine so that both exhaust pushrods are at the bottom of their stroke and follow an identical procedure for the exhaust rocker box.

3 Tighten down all the rocker box and cylinder head retaining bolts, following the sequence shown in the accompanying illustration. Turn each bolt, one turn at a time, before proceeding to the next in sequence. Use a torque wrench to achieve the final settings of 18 ft lb for the 3/8 in diameter bolts and 15 ft lb for the 5/16 in diameter bolts. Note that if the cylinder head is not tightened down in the correct sequence, there is every possibility that it will be distorted permanently.

36 Adjusting the valve clearance

1 Before refitting the inspection caps or finned covers to the rocker boxes, it is necessary to check, and if necessary reset, the valve clearances. To check the left hand valve clearance, turn the engine over until the right hand exhaust valve is fully open. This will ensure that the left hand tappet is resting on the base circle of the cam. Some cams have so-called quietening ramps and this procedure affords the only reliable means of ensuring the tappet follower is positioned at minimum lift.

2 The exhaust valve clearance is 0.004 in set with the engine cold. A feeler gauge of this thickness should be a good sliding fit between the end of the valve stem and the rocker arm if the adjustment is correct. To reset the gap, slacken the locknut of the adjuster screw at the rocker arm tip and obtain the correct gap by varying the setting of the adjuster screw. When the gap is correct, tighten the locknut and recheck. Repeat this procedure for the right hand exhaust valve, rotating the engine so that the left hand exhaust valve is fully open. Replace both inspection caps, using new sealing washers. It may be necessary to slacken the crosshead screws to release the retaining clips holding each inspection cap in place, before the caps can be screwed home fully. When the caps have been tightened, retighten the crosshead screws and check that each retaining clip bears on the milled edge of its respective inspection cap. Post 1970 models use a different arrangement, in the form of finned covers attached by bolts.

3 The inlet valve clearance is 0.002 in, set with the engine cold. Use an identical procedure to that adopted for the exhaust valve settings; when one valve clearance is being checked the corresponding valve in the other cylinder should be on full lift. Replace the inlet rockerbox inspection caps, or covers.

4 It is not easy to insert feeler gauges when checking the tappet clearances and hence it is often of help to know that a quarter turn (90°) of each adjuster will result in a difference of 0.010 inch. From the fully closed position, (no clearance) a 40° turn will give a 0.004 inch clearance and a 20° turn a clearance of 0.002 inch. Always re-check after tightening the lock nut, and if necessary, re-adjust.

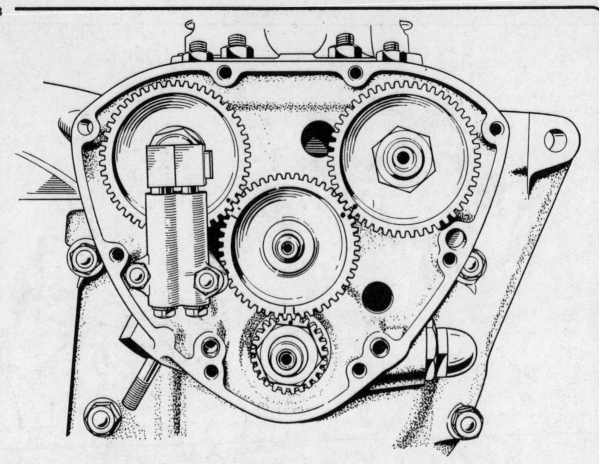

FIG.1.9. INTERMEDIATE WHEEL LOCATION

Exhaust camshaft pinion dot aligned with dot on intermediate wheel

Crankshaft pinion dot aligned with twin dashes on intermediate wheel

Inlet crankshaft pinion. Dot aligned with:- long dash for T120 and TR6, short dash for 6T.

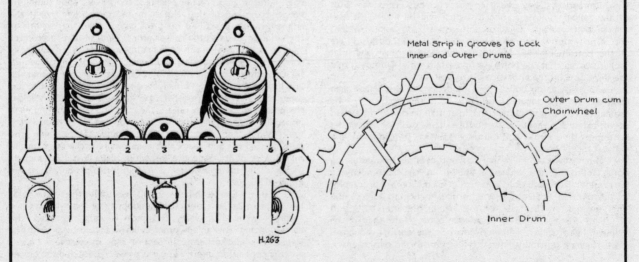

H.263

FIG.1.10. Aligning pushrod etc   FIG. 1.11. Method of locking clutch inner and outer drums

34.1a Position engine unit in near correct position first

34.1b Line up bolt holes before inserting front mounting bolt first

35.1a Insert pushrods. Note identification marking

35.1b Use new gaskets for rocker boxes to prevent oil leaks

36.2a Check valve clearances with feeler gauge

36.2b To adjust, slacken locknut and turn squared end

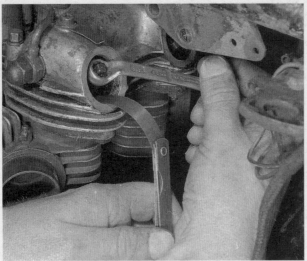

36.2c Always recheck after tightening locknut

36.2d Tighten inspection caps fully to prevent loss

37.1 Head steadies must be tight to prevent vibration

37.3 Take care rocker feed pipe does not twist whilst tightening nuts

38.1 Coils must face rearwards when refitted

38.3a If top is fitted first, carburettor replacement is easier

37 Engine reassembly - fitting the head steadies and rocker oil feed pipe

1 Replace the four short head steady stays, complete with their distance pieces. All nuts and bolts must be tightened fully when they are located correctly; a loose head steady assembly will accentuate engine vibration.

2 Anneal the four copper sealing washers that fit over the rocker spindle unions by heating them to red heat, then plunging them into cold water. Remove any scale which may have formed during this operation. Annealing makes the washers more malleable and therefore better able to form an effective oil seal.

3 Place one washer over each rocker spindle extension, then fit the feed pipe. Fit another washer on the outside of each feed pipe union, then the domed nuts. Take care when tightening the domed nuts to ensure the very thin feed pipe does not turn and 'neck'. If the pipe bend is too sharp or if the pipe is restricted at any point, the rate of oil flow to the rocker gear will be restricted.

4 Check that the rubber joint at the extreme end of the rocker feed pipe is in good condition and makes good connection with the T piece in the oil return pipe to the oil tank.

38 Engine reassembly - completion and final adjustments

1 Refit the coils to the lug immediately above the exhaust rocker boxes. The closed end of each coil should face the front of the machine, to prevent the ingress of water. Replace the colour-coded wires by means of their snap connectors and reconnect the spark plug leads with their respective spark plugs.

2 Rejoin the electrical connections from the generator, not forgetting those below the crankcase. Replace the fuse in the fuse holder close to the battery, or the battery leads, if they were detached.

3 Replace the carburettor(s) using a new gasket between the heat insulating distance piece and the flange of the cylinder head. Insert the slide and needle assembly, making sure the needle locates with the needle jet, before the top of the carburettor(s) is tightened down. Fitting is easier if top is assembled to carburettor(s) first. Replace the exhaust system and tighten the finned clamps at the cylinder head junction.

4 Before replacing the petrol tank, make sure the throttle and air cables are routed so that they will not be trapped or bent into tight curves. Heavy throttle action or a tendency for the slide(s) to stick can often be traced to poor cable routing.

5 The petrol tank is retained by two rubber mountings at the front and a nut and washer which compress these mountings, when tightened. It is advisable to drill and wire these two nuts together as a safeguard against them slackening off. The rear mounting takes the form of a single bolt and washer that passes through a rubber on each side of the tank lug. The bolt threads into the frame tube. Connect the petrol pipes to each petrol cap.

6 Replace the two footrests and the rear brake pedal. The footrests are secured by a nut and washer on the inside of each rear engine plate. Two locating pegs prevent the footrests from rotating if the nut should slacken. The brake pedal passes through a short tube welded to the left hand rear engine plate and has a square end which locates with the operating arm connected to the rear brake rod.

7 Reconnect the final drive chain. It is easiest to insert the spring link if the ends of the chain are pressed into the rear wheel sprocket. Make sure that the closed end of the link faces the direction of chain travel.

8 Connect the oil pipes from the engine to the oil tank by the flexible hoses provided. The hoses should be secured firmly with the screw clamps at each end. Note that the forward facing of the two pipes from the engine is the feed pipe and must be connected to the rearmost of the pipes from the base of the oil tank which has a union connection. The other pipe is the scavenge or oil return pipe. When the pipes are connected, refill the oil tank with 6 pints of the recommended grade of engine oil.

9 Check the oil level in the gearbox and if necessary top up. Reconnect the clutch cable and adjust so that there is only a small amount of play at the handlebar lever before the clutch commences to operate. It is possible that the clutch action may be either too heavy or too light, which is why the outer chaincase has been attached only temporarily. If adjustment of the clutch springs is required, the outer cover must be removed to gain access, also the left hand footrest. When the adjustment appears correct, check that the clutch does not slip when the engine is turned over against compression. Check that it will slip freely when the clutch is withdrawn fully and that the pressure plate spins freely, without wobble. Wobble can be corrected by adjusting the tension of the clutch springs individually. Replace the outer chaincase, using a new gasket and a generous smear of gasket cement to both jointing faces. Tighten the cover fully before replacing the left hand footrest. Finally, adjust the primary chain tension until there is about ½ in play in the middle of the chain run, as felt through the inspection cap which acts as a filler plug. Replace the drain plug that blanks off access to the chain tensioner adjuster in the base of the chaincase and add 5/8 pint of engine oil.

10 Reconnect the speedometer and tachometer drives. Connect the electrical lead to the oil pressure switch, at the front of the timing chest on all post 1968 models.

39 Starting and running the rebuilt engine

1 Switch on the ignition and run the engine slowly for the first few minutes, especially if the engine has been rebored. Remove the cap from the top of the oil tank and check that oil is returning. There may be some initial delay whilst the pressure builds up and oil circulates throughout the system, but if none appears after the first few minutes running, stop the engine and investigate the cause. If the pressure release valve is unscrewed a few threads, oil should ooze from the joint if the oil pump is building up pressure.

2 Check that all controls function correctly and that the generator is indicating a charge on the ammeter. Check for any oil leaks or blowing gaskets.

3 Before taking the machine on the road, check that all the legal requirements are fulfilled and that items such as the horn, speedometer and lighting equipment are in full working order. Remember that if a number of new parts have been fitted, some running-in will be necessary. If the overhaul has included a rebore, the running-in period must be extended to at least 500 miles, making maximum use of the gearbox so that the engine runs on a light load. Speeds can be worked up gradually until full performance is obtainable by the time the running-in period is completed.

4 Do not tamper with the exhaust system under the mistaken belief that removal of the baffles or replacement with a quite different type of silencer will give a significant gain in performance. Although a changed exhaust note may give the illusion of greater speed, in a great many cases quite the reverse occurs in practice. It is therefore best to adhere to the manufacturer's specification.

40 Engine modifications and tuning

1 The Triumph twin engine can be tuned to give even higher performance and yet retain a good standard of mechanical reliability. Many special parts for boosting engine performance are available both from the manufacturer and from a number of specialists who have wide experience of the Triumph marque. The parts available include high compression pistons, high lift camshafts and even cylinder heads with four valves per combustion chamber.

2 There are several publications, including a pamphlet available from the manufacturer, that provide detailed information about the ways in which a Triumph twin engine can be modified to give increased power output. It should be emphasised, however, that a certain amount of mechanical skill and experience is

necessary if an engine is to be developed in this manner and still retain a good standard of mechanical reliability. Often it is preferable to entrust this type of work to an acknowledged specialist and therefore obtain the benefit of his experience.

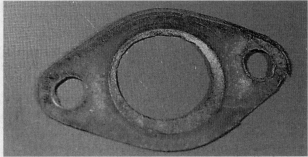

38.3b. How to lose performance. This gasket was too small

38.3c Fit silencers after exhaust pipes are in position

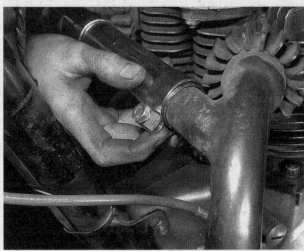

38.3d Do not forget to tighten clamps on balance pipe

38.6a Replace footrests so that pegs locate with holes in engine plates

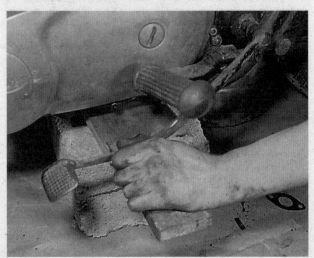

38.6b Brake pedal passes through tube welded to left hand engine plate

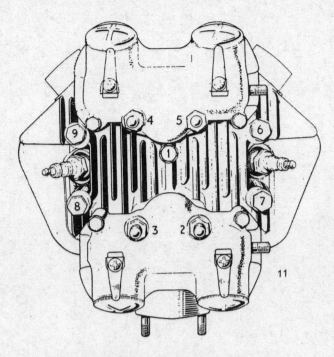

Fig. 1.12 Cylinder head bolt tightening sequence

41 Fault diagnosis

Symptom	Reason/s	Remedy
Engine will not turn over	Clutch slip	Check and adjust clutch.
	Mechanical damage	Check whether valves are operating correctly and dismantle if necessary.
Engine turns over but will not start	No spark at plugs	Remove plugs and check. Check whether battery is discharged.
	No fuel reaching engine	Check fuel system.
	Too much fuel reaching engine	Check fuel system. Remove plugs and turn engine over several times before replacing.
Engine fires but runs unevenly	Ignition and/or fuel system fault	Check systems as though engine will not start.
	Incorrect valve clearances	Check and reset.
	Burnt or sticking valves	Check for loss of compression.
	Blowing cylinder head gasket	See above.
Lack of power	Incorrect ignition timing	Check accuracy of setting.
	Valve timing not correct	Check timing mark alignment on timing pinions.
	Badly worn cylinder barrel and pistons	Fit new rings and pistons after rebore.
High oil consumption	Oil leaks from engine gear unit	Trace source of leak and rectify.
	Worn cylinder bores	See above.
	Worn valve guides	Replace guides.
Excessive mechanical noise	Failure of lubrication system	Stop engine and do not run until fault located and rectified.
	Incorrect valve clearances	Check and re-adjust.
	Worn cylinder barrel (piston slap)	Rebore and fit oversize pistons.
	Worn big end bearings (knock)	Fit new bearing shells.
	Worn main bearings (rumble)	Fit new journal bearings.

Chapter 2 Gearbox

Contents

Specifications

Overall gear ratios

	4 speed gearbox	5 speed gearbox	
		650 cc models	750 cc models
Fifth gear	—	4.95	4.7
Fourth gear	5.84	5.89	5.59
Third gear	5.76	6.92	6.58
Second gear	8.17	9.07	8.63
First gear	11.8	12.78	12.25

Bearings

Mainshaft (right)	¾ in x 1⁷/8 in x 9/16 in ball journal
Sleeve gear (left)	1¼ in x 2½ in x 5/8 in ball journal
Layshaft (right)	11/16 in x 7/8 in x ¾ in needle roller
(left)	11/16 in x 7/8 in x ¾ in needle roller

Springs (free length)

Camplate plunger	2½ in
Gearchange plunger	1¼ in each
Gear lever return	1¾ in each
Kickstarter ratchet	½ in

Gearbox sprocket ... ...

19 teeth (solo)   20 teeth (solo) **750 cc models**
17 teeth (sidecar)

1 General description

A four-speed constant mesh gearbox is fitted in unit with the engine on all 650 cc unit-construction models. The 650 cc Bonneville model, the T130V, is also available, with a five-speed gearbox. Both versions of the 750 cc model, the TR7V Tiger 750 and the T140V Bonneville 750 are fitted with five-speed gearboxes as standard. Apart from the addition of the extra gear pinions, the four and five-speed gearboxes are similar in most respects.

It is not necessary to dismantle the engine in order to gain access to the gearbox, although if a complete overhaul is necessary it may prove more convenient to lift the engine unit from the frame in order to gain better access. Unlike most other British manufacturers, Triumph Engineering have remained faithful to their 'up for up' and 'down for down' method of gear selection, which can be changed to the more conventional

arrangement only if a reverse camplate is fitted or the gear change pedal is reversed on its shaft so that it faces rearward.

2 Dismantling the gearbox - general

1   Before commencing work on the gearbox, make sure that the outer surfaces of the gearbox end covers and shell are clean and dry. Lay a sheet of clean paper immediately below the gearbox so that any parts inadvertently misplaced will fall onto a clean surface.

2   A good fitting crosshead screwdriver is essential to prevent damage to the five screws retaining the outer end cover in position; also a slim socket or box spanner to fit the recessed nuts which complete the method of end cover retention. If the screw heads are damaged, it will prove difficult to remove or replace the screws, even when the correct screwdriver bit is available.

3.2 Detach cable nipple from operating arm with small screw-driver

4.2 Remove ratchet retaining nut, then ratchet assembly

3 Dismantling the gearbox - removing the outer end cover

1   Before access can be gained to the outer end cover, it is necessary to remove the right hand exhaust pipe, silencer and the right hand footrest. Slacken the finned clip fitted to the right hand exhaust pipe and remove the bolts from the exhaust pipe stay and from the lug of the silencer. If the exhaust system is 'siamesed' or has a balance pipe between the right and left hand exhaust pipes, slacken the clamp around the joint between both systems. The exhaust pipe can now be freed by a few taps with a rawhide mallet and the complete right hand system removed. The footrest is secured by a nut and washer on the inside of the rear engine plate.

2   Slacken off the clutch cable adjuster at the handlebar lever and withdraw the cable and nipple. Remove the slotted plug from the gearbox outer cover and detach the other end of the clutch cable from the operating lever by displacing the nipple with a screwdriver. The clutch cable can now be removed completely by sliding up the rubber cover over the adjuster threaded into the top of the end cover casing and unscrewing the adjuster, after slackening the locknut.

3   Some of the earlier models do not have a slotted plug in the gearbox end cover. In these circumstances, the slotted adaptor should be removed and the gearbox end cover adjuster unscrewed to expose the nipple at the gearbox end of the clutch cable. This nipple engages with a slotted barrel-shaped connector attached to the clutch operating arm and can be slid out of position when the cable is pulled upwards.

4   Place a tray beneath the gearbox and unscrew the gearbox drain plug. The gearbox holds just under one pint of oil.

5   Engage top gear. Although not strictly necessary, it will facilitate the removal of several nuts which have to be removed during the dismantling sequences to follow by permitting the engine to be locked, when the rear brake is applied.

6   Unscrew the top and bottom hexagonal nuts and the five crosshead screws from the end cover. Depress the kickstarter slightly, to clear it from the top, and tap the end cover until it can be drawn away from the gearbox shell.

7   There is no necessity to remove either the gear change lever or the kickstarter unless attention is required to either mechanisms.

4 Dismantling the gearbox - removing the inner end cover and gear clusters

1   Removal of the inner cover may be hampered by the flexible oil pipes immediately below the gearbox. If it is necessary to remove them, the oil tank must first be drained. It is recommended that the pipes are detached by removing the union at the point where the pipes join the crankcase, which is more accessible after the gearbox outer cover has been removed. The union is retained to a stud by one centre nut and washer; there is no chance of the pipes being inadvertently reversed if this procedure is adopted.

2   Before the inner cover can be removed, it will be necessary to detach completely the rear right hand engine plate which forms part of the mounting for the unit-construction engine. Remove the kickstarter pinion ratchet retaining nut from the end of the gearbox mainshaft, after bending back the tab washer. It can be locked by applying the back brake whilst the gearbox is held in top gear. Draw the ratchet assembly off the mainshaft complete with spring and inner bush. Withdraw the clutch pushrod from the centre of the mainshaft.

3   Machines having an engine number prior to DU 24875 have the speedometer drive taken from the gearbox. It will be necessary to release the speedometer drive cable by unscrewing the union nut so that the cable is withdrawn before the inner end cover is detached.

3   If it is necessary to remove the final drive sprocket, to renew the gearbox main bearing or oil seal, it is advisable to remove the outer chaincase cover at this stage and dismantle the primary drive, as described in Chapter 1, section 10. It is much more difficult to remove these components at a later stage since there will be no support for the gearbox mainshaft and layshaft after the inner cover is removed.

4   Unscrew the large domed nut from beneath the gearbox and withdraw it complete with the camplate plunger and spring. Then remove the gearbox inner cover by unscrewing the crosshead screw, Allen screw and bolt that retains it in position. The latter is found within the recessed portion of the end cover, close to the oil pipe union. A few light taps with a rawhide mallet may be necessary to displace the cover so that it can be slid off, over the end retaining studs.

5   Remove the selector fork spindle, which is a push fit into the gearbox shell. The gearbox mainshaft can now be withdrawn together with the lower gear pinions and the selectors. The layshaft and the remaining gear pinions can be withdrawn, leaving only the camplate assembly and the mainshaft top gear. The camplate will pull off its shaft without difficulty. Do not lose the two brass thrust washers which locate with the needle roller bearings or the rollers from the selector forks.

6   To remove the mainshaft top gear (sleeve gear pinion) it is necessary to remove the circular cover from the chaincase behind the clutch and bend back the tab washer before unscrewing the large retaining nut (right hand thread). The nut measures 1.66 in across the flats and can be removed by Triumph service tool Z63 if a spanner of a suitable size is not available. When the nut is removed, the sleeve gear pinion is released by driving it into the gearbox shell with a hammer and soft metal drift. After the chain has been detached this action will free the final drive sprocket from the splines and permit the sprocket to be passed through the aperture in the chaincase.

4.4 Inner cover is retained by two screws and a bolt

4.5a Selector fork spindle is push fit in gearbox shell

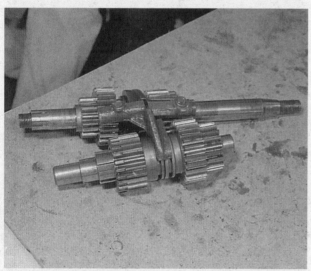
4.5b The gear pinions and selectors after removal from shell

4.6a Sleeve gear pinion passes through left hand ball race

4.6b Remove cover in rear of chaincase for access to sleeve gear pinion

4.6c Sprocket is released after tab washer and nut are removed

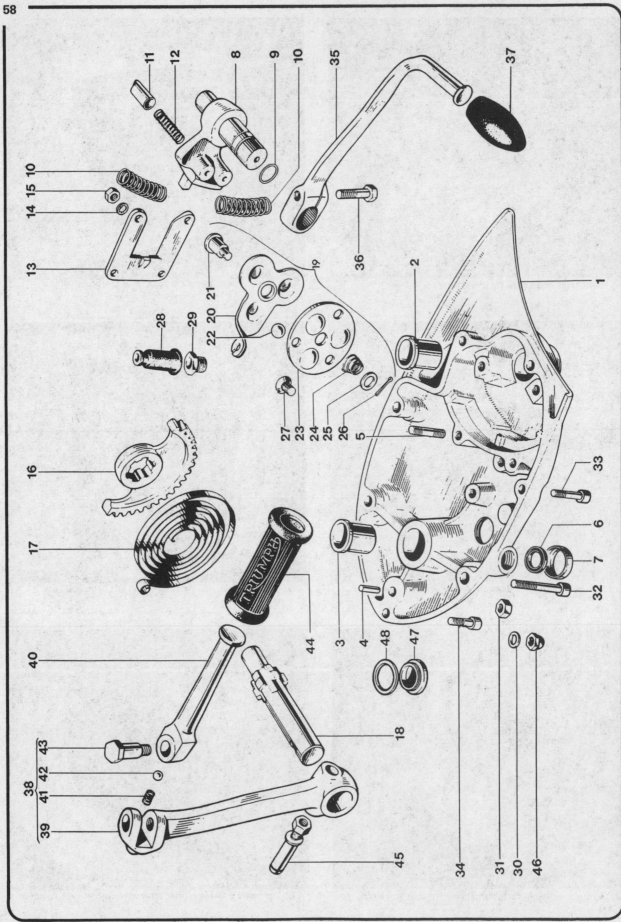

FIG. 2.1. GEARBOX OUTER COVER – COMPONENTS

1 Gearbox outer cover
2 Gearchange spindle bush
3 Kickstarter spindle bush
4 Kickstarter anchor pin
5 Guide plate stud - 4 off
6 Oil seal
7 Oil seal housing
8 Gear change quadrant
9 Rubber 'O' ring
10 Quadrant return spring - 2 off
11 Gearchange quadrant plunger - 2 off
12 Plunger spring - 2 off

13 Guide plate
14 Serrated washer - 4 off
15 Nut - 4 off
16 Kickstarter quadrant
17 Kickstarter return spring
18 Kickstarter spindle
19 Clutch operating arm assembly
20 Clutch operating arm
21 Pivot for operating arm
22 Ball bearing, 3/8 inch diameter - 3 off
23 Thrust plate
24 Return spring

25 Plain washer
26 Split pin
27 Countersunk screw - 2 off
28 Rubber cover
29 Abutment for clutch cable
30 Plain washer - 2 off
31 Nut
32 Screw - 2 off
33 Screw - 2 off
34 Screw
35 Gearchange lever
36 Bolt

37 Rubber for gearchange lever
38 Folding kickstarter
39 Kickstarter lever
40 Kickstarter pedal
41 Pedal locating spring
42 Steel ball, ¼ inch diameter
43 Pedal bolt
44 Kickstarter rubber
45 Cotter pin, with nut and washer
46 Domed nut
47 Filler plug
48 Fibre washer

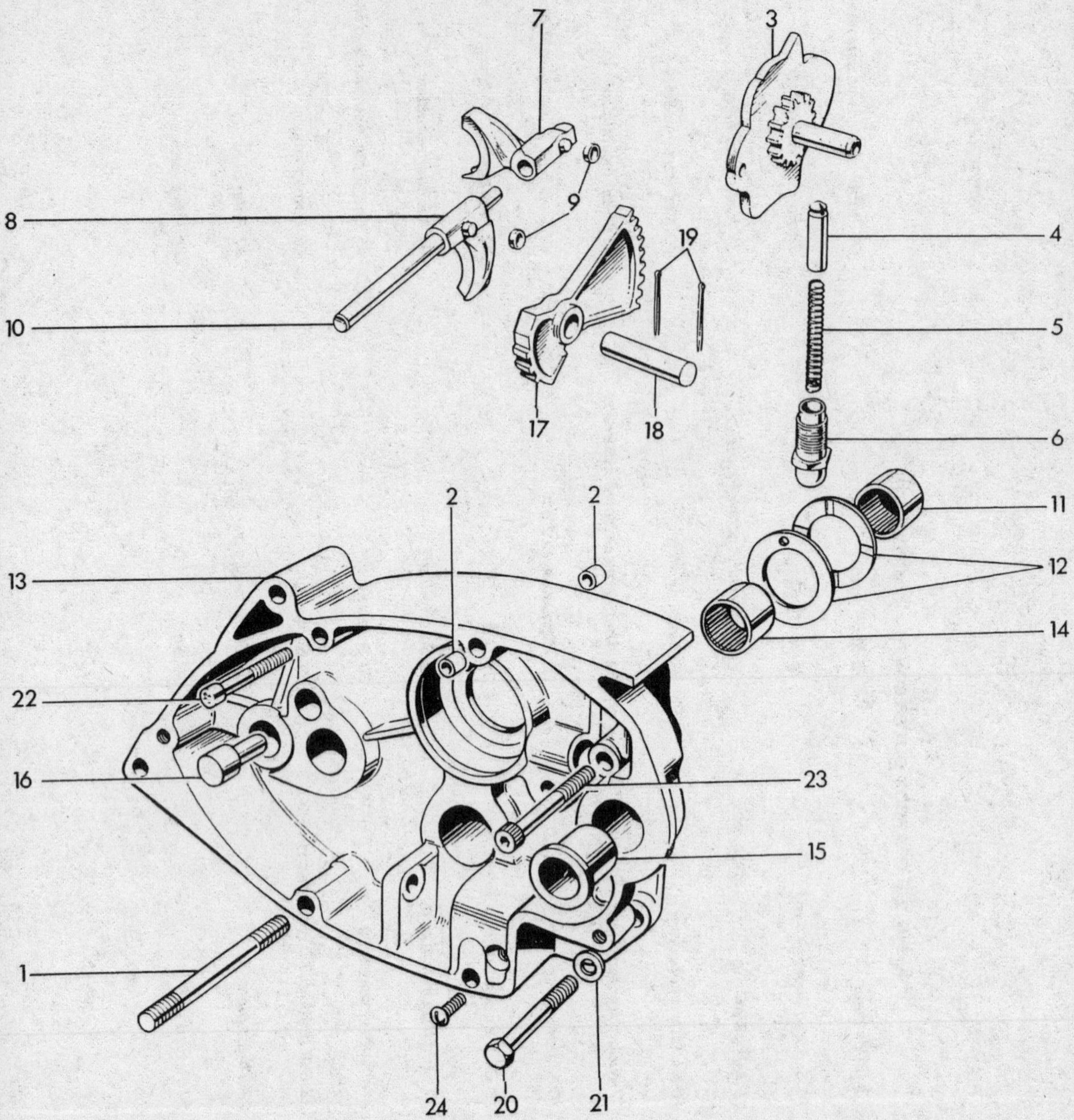

7

3

8

10

4

5

6

19

17   18

2   2

13

11

12

22

14

16

23

15

1

24   20   21

FIG. 2.2. GEARBOX INNER COVER AND SELECTORS

1	Stud - 2 off	13	Gearbox inner cover
2	Hollow dowel - 4 off	14	Layshaft bearing
3	Gear selector camplate	15	Gearchange spindle bush
4	Camplate index plunger	16	Kickstarter stop
5	Index plunger spring	17	Camplate operating quadrant
6	Index plunger holder	18	Quadrant spindle
7	Mainshaft selector fork	19	Split pin - 2 off
8	Layshaft selector fork	20	Bolt
9	Selector fork roller - 2 off	21	Plain washer
10	Selector fork rod	22	Screw
11	Layshaft bearing	23	Socket screw
12	Thrust washer - 2 off	24	Countersunk screw

5 Dismantling the gearbox - removing (and replacing) the mainshaft and layshaft bearings

1 The left hand sleeve gear bearing is retained in the gearbox shell by a circlip which is preceded by an oil seal. Since it is necessary to renew the oil seal whenever the bearing is disturbed, prise the old seal out of position and remove the circlip with a pair of long nosed pliers.

2 Heat the gearbox shell locally with a blow lamp and drive the bearing out from the inside of the gearbox using a suitable drift. Before the shell cools, fit the replacement bearing and make sure it is pressed to the full depth of the housing before the circlip is fitted. Fit the new oil seal, after the shell has cooled.

3 The right hand mainshaft bearing is also retained by a circlip in the gearbox inner end cover. Heat treatment will again be required to help free the bearing after the circlip has been removed. The replacement bearing should be located before the end cover is allowed to cool and the circlip replaced.

4 The caged needle roller bearings used to support each end of the layshaft are removed and replaced in similar fashion. It is, however, necessary to ensure that the new bearings are pressed in, plain end first, whilst the gearbox shell or cover is still warm and allowed to protrude 0.073 in to 0.078 in INSIDE the gearbox. This protruding lip forms the anchorage for the two brass thrust washers mentioned in Section 4.5. In each case, the outer face of the bore into which the needle roller bearings fit should be sealed with a proprietary sealant to obviate the risk of oil leakage.

6 Examination and renovation - general

1 Before the gearbox is reassembled, it will be necessary to inspect each of the components for signs of wear or damage. Each part should be washed in a petrol/paraffin mix to remove all traces of oil and metallic particles which may have accumulated as the result of general wear and tear within the gearbox.

2 Do not omit to check the castings for cracks or other signs of damage. Small cracks can often be repaired by welding, but this form of reclamation requires specialist attention. Where more extensive damage has occurred it will probably be cheaper to purchase a new component or to obtain a serviceable second-hand part from a breaker.

3 If there is any doubt about the condition of a part examined, especially a bearing, it is wise to play safe and renew. A considerable amount of strip down work will be required again if the part concerned fails at a much earlier date than anticipated.

7 Kickstarter mechanism - examination and renovation

1 If the kickstarter quadrant shows signs of wear, it should be renewed, otherwise it will tend to jam during initial engagement with the ratchet pinion. Note that the first tooth of the quadrant is relieved to minimise the risk of a jam when the initial engagement is made.

2 The quadrant is a drive fit on the splines of the kickstarter shaft. The kickstarter is removed by unscrewing the nut on the end of the cotter pin and driving out the pin with a hammer. When replacing a quadrant, it is important to ensure that the flat in the shaft is positioned correctly in relation to the quadrant or the operating angle will be incorrect. A new oil seal should be fitted over the shaft in the outer face of the end cover.

3 Renew the return spring if the action is weak or if the spring has suffered damage. The accompanying diagram shows the correct location of the spring prior to tensioning.

4 If the kickstarter ratchet has to be renewed, it is probable that the ratchet pinion, with which it engages, will require renewal too. It is bad policy to run old and new parts together.

5 It is also important to check the condition of the ratchet teeth and to renew both parts of the ratchet system if the teeth are rounded at their edges. A worn ratchet will eventually slip and make starting difficult. Renew the light return spring if it has taken a permanent set or has reduced return action.

5.1 Sleeve gear pinion bearing is retained by circlip

5.2 Warm gearbox shell before bearing is knocked out

7.1 Kickstarter quadrant has the leading tooth relieved

8 Gearchange mechanism - examination and renovation

1 It should not be necessary to dismantle the gear change mechanism unless the gear lever return springs have broken or wear of the operating mechanism is suspected on account of imprecise gear changes. To dismantle the assembly, detach the gear change lever by slackening the pinch bolt and pulling the lever off the splined shaft. Remove the four nuts and lockwashers securing the guide plate to the inside of the outer end cover and withdraw the guide plate complete with plunger quadrant and the curved gear change lever return springs.

2 Examine the various components for wear, especially the gear change plungers and the plunger springs. Each spring should have a free length of 1¼ in; renew them if they have taken a set. The plungers must be a clearance fit in the quadrant if they are to function correctly.

3 If the plunger guide plate is worn or grooved on the taper guide surfaces it must be renewed.

4 The gear change lever return springs seldom give trouble unless they become fatigued or if condensation within the gearbox causes corrosion. If there is any doubt about their condition renew them as a matter of course.

5 After a lengthy period of service the gear change quadrant bush may wear oval. If this form of wear is evident, the bush must be renewed.

6 If the teeth of the camplate operating quadrant attached to the inner end cover are chipped, indented or worn, the quadrant must be renewed. It is retained by two split pins which, when removed, will permit the spindle to be withdrawn.

7 Do not omit to check the oil seal around the shaft of the gear change lever in the outer end cover.

9 Clutch operating mechanism - examination and renovation

1 The clutch operating mechanism, attached to the outer end cover by two screws, seldom gives trouble provided the gearbox oil level is maintained.

2 The component parts are assembled as shown in Fig 3.1 of Chapter 3.

10 Gearbox components - examination and renovation

1 Examine each of the gear pinions carefully for chipped or broken teeth. Check the internal splines and bushes. Instances have occurred where the bushes have worked loose or where the splines have commenced to bind on their shafts. The two main causes of gearbox troubles are running with low oil, and condensation, which gives rise to corrosion. The latter is immediately evident when the gearbox is dismantled.

2 The mainshaft and the layshaft should both be examined for fatigue cracks, worn splines or damaged threads. If either of the shafts have shown a tendency to seize, discoloration of the areas involved should be evident. Under these circumstances check the shafts for straightness.

3 Harsh transmission is often caused by rough running ball races especially the mainshaft ball journal bearings. Section 5 of this Chapter describes the procedure for removing and replacing the gearbox bearings.

4 All gearbox bearings should be a tight fit in their housings. If a bearing has worked loose and has revolved in the housing, a bearing sealant such as Loctite can be used, provided the amount of wear is not too great.

5 Check that the selector forks have not worn on the faces which engage with the gear pinions and that the selector fork rod is a good fit in the gearbox housings. Heavy wear of the selector forks is most likely to occur if replacement of the mainshaft bearings is long overdue.

6 The gear selector camplate will wear rapidly in the roller tracks if the mainshaft bearings need replacement. Although the gear pinion behind the camplate is unlikely to wear excessively, it should be inspected if it has proved difficult to select gears.

7 The camplate plunger must work freely within its housing.

Check the free length of the spring, which should be 2½ in if it has not compressed.

8 Play, accompanied by oil leakage, is liable to occur if the bush within the sleeve gear pinion is worn. The working clearance is normally from 0.003 in to 0.005 in.

11 Gearbox reassembly - general

1 Before commencing reassembly, check that the various jointing surfaces are clean and undamaged and that no traces of old gasket cement remain. This check is particularly important because the gearbox itself is assembled with face to face joints and no gaskets. The gearbox will not remain oiltight if these simple precautions are ignored.

2 Check that all threads are in good condition and that the locating dowels, where fitted, are positioned correctly. Have available an oil can filled with clean engine oil so that the various components can be lubricated during reassembly.

12 Gearbox reassembly - replacing the sleeve gear pinion and final drive sprocket

1 Assuming the mainshaft main bearing is in position in the extreme left hand end of the gearbox shell, and retained by a circlip, drive the new oil seal into position with the lip and spring innermost. Lubricate the sleeve gear pinion and drive it through the main bearing from inside the gearbox, taking special care that the feather edge of the new oil seal is not damaged.

2 Lubricate the ground, tapered boss of the final drive sprocket with oil and slide it on to the splines of the sleeve gear. Replace the tab washer and then the large retaining nut which should be fingertight at this stage. Reconnect the final drive chain (closed end of the spring link facing the direction of travel of the chain) and apply the rear brake so that the sprocket retaining nut can be tightened fully. Bend over the tab washer so that the nut is retained in place.

13 Gearbox reassembly - replacing the camplate, camplate plunger, gear pinions and selectors

1 Lubricate the camplate spindle and replace the camplate in its correct location within the gearbox. Assemble the camplate plunger and spring into the domed retaining nut, and check the plunger moves quite freely when it is within its housing. Place a new fibre washer above the domed nut and screw the plunger assembly into the bottom of the gearbox. Tighten the nut and check that the camplate is positioned so that the plunger engages with the depression BETWEEN second and third gear notches. This position must be maintained whilst the gearbox is re-assembled, otherwise the gears will not index correctly.

2 Locate one of the brass thrust washers over the layshaft inner needle bearing. The oilways should face outward, towards the open end of the gearbox. Grease can be used to hold this washer in place whilst reassembly continues.

3 Lubricate the gear pinions attached to the mainshaft and layshaft, then assemble the mainshaft and layshaft gear clusters. Grease the camplate rollers to retain them on the selector forks and position the forks. Note that the fork with the smaller radius must engage with the mainshaft gear cluster.

4 Insert the gear train and selectors as a complete unit and ensure that the camplate rollers engage correctly with the roller track in the camplate. This is a somewhat tricky operation and it may be necessary to make several attempts before all of the components locate correctly.

5 When assembly is complete, lubricate the selector fork spindle and insert it through the selector forks after checking that the bores are approximately in line. The shouldered end of the rod should be inserted first and the shaft pushed home until it engages fully with the gearbox housing. Check that the mainshaft selector fork is in the innermost position.

12.2 Lock sprocket nut with tab washer, after tightening fully

13.1a Insert camplate plunger and spring

13.1b Camplate MUST be in position shown throughout assembly

14.2a Hold camplate quadrant in correct position when pushing cover home

14.2b Camplate quadrant must align as shown for correct gear indexing

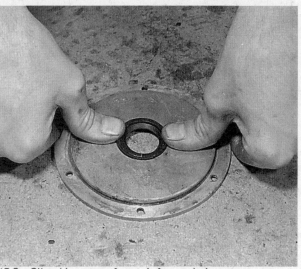
15.3a Oil seal in centre of cover before replacing cover

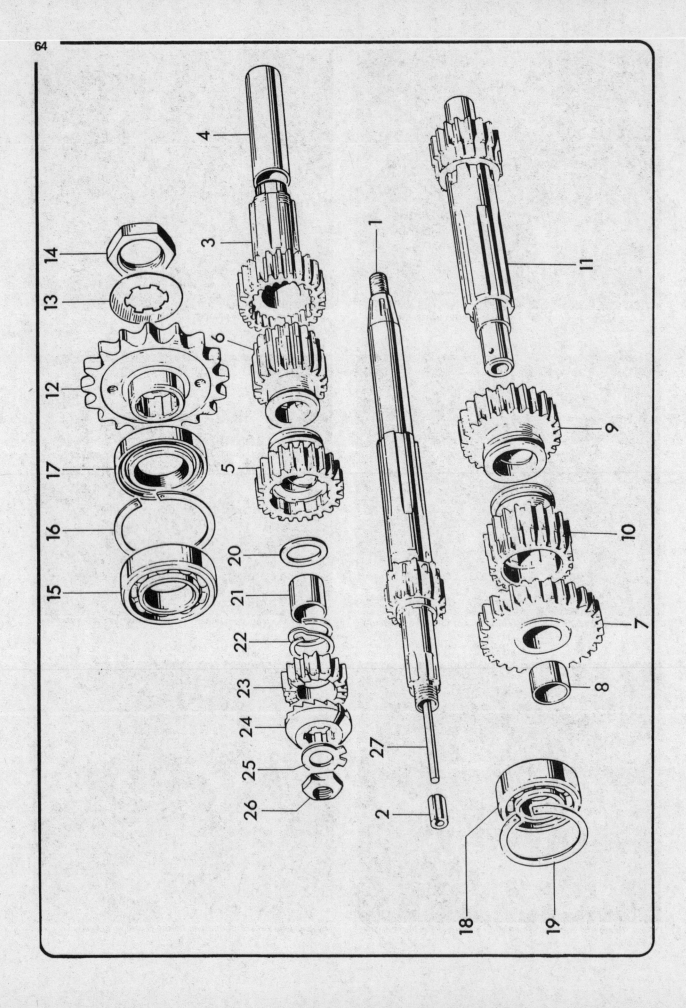

FIG. 2.3. FOUR-SPEED GEARBOX: SHAFTS AND GEAR PINIONS

1	Mainshaft complete with bottom gear pinion (16 teeth)	10	Layshaft third gear pinion (22 teeth)
2	Clutch rod bush	11	Layshaft complete with top gear pinion
3	Mainshaft top gear (sleeve) pinion (26 teeth)	12	Gearbox sprocket (available with 15 - 20 teeth)
4	To gear bush	13	Tab washer
5	Mainshaft third gear pinion (24 teeth)	14	Gearbox sprocket nut
6	Mainshaft second gear pinion (20 teeth)	15	Top gear (sleeve) bearing
7	Layshaft bottom gear pinion (30 teeth)	16	Circlip
8	Bottom gear bush	17	Oil seal
9	Layshaft second gear pinion (26 teeth)	18	Mainshaft bearing
		19	Circlip
		20	Plain washer
		21	Kickstarter pinion sleeve bush
		22	Kickstarter pinion spring
		23	Kickstarter pinion
		24	Kickstarter ratchet
		25	Tab washer
		26	Nut
		27	Clutch operating rod

15.3b Take care not to damage seal when positioning over gearbox mainshaft

15.5 Tighten clamp to prevent lever working loose on splines

6   Place the other brass thrust washer over the end of the needle roller bearing of the inner end cover and check that the camplate operating lever moves freely. Grease can again be used to retain the thrust washer during assembly.

14   Gearbox reassembly - replacing the inner end cover and indexing the gears

1   Use the oil can to lubricate all parts at present located within the gearbox shell. It is advisable to replace the gearbox drain and level plug first to prevent subsequent leakage.
2   Apply a thin coating of gasket cement to the face joint of the gearbox shell and to the back of the inner end cover which is to mate with it. Replace the two dowels at the top and bottom of the gearbox shell face joint and check that the camplate is still positioned so that the plunger is in full engagement with the depression between the second and third gear notches. Slide the inner end cover over the two projecting studs of the gearbox shell and, when it is only about ¼ inch from the jointing face, position the camplate quadrant so that it is exactly in position. Push the cover fully home and again check that the alignment of the camplate quadrant is correct. The need for careful alignment cannot be overstressed since this determines the correct indexing of the gears. If the alignment is as much as

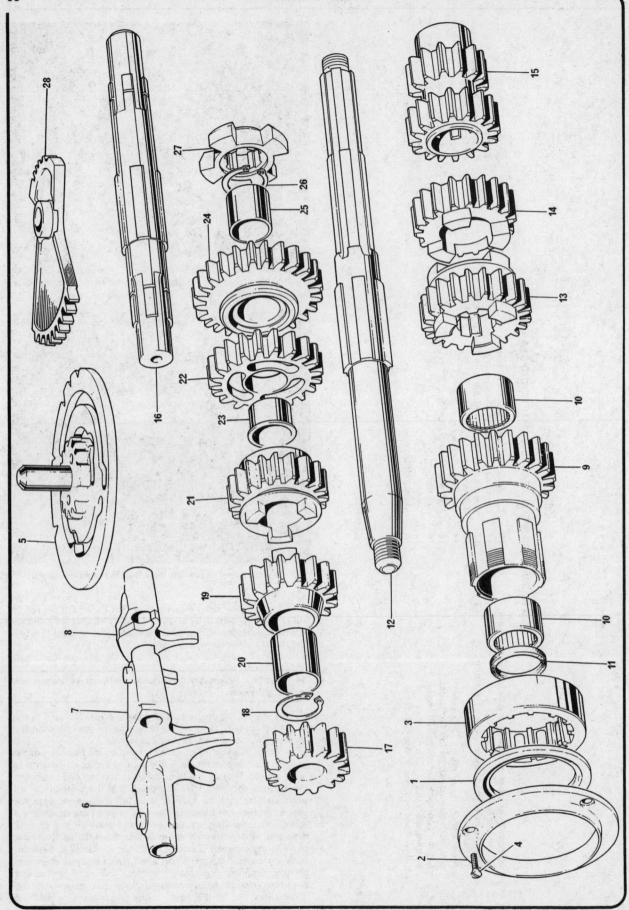

FIG. 2.4. FIVE-SPEED GEARBOX - AND GEAR PINIONS

1 Oil seal
2 Oil seal housing
3 Top gear bearing
4 Screw for oil seal housing - 3 off
5 Gearbox camplate
6 Mainshaft 4th gear selector fork
7 Layshaft 3rd gear selector fork
8 Layshaft bottom gear selector fork
9 Mainshaft top gear pinion
10 Needle roller bearing - 2 off
11 Oil seal
12 Mainshaft
13 Mainshaft 4th gear pinion
14 Mainshaft 3rd gear pinion
15 Mainshaft bottom and 2nd gear pinions
16 Layshaft
17 Layshaft top gear pinion
18 Circlip
19 Layshaft 4th gear pinion
20 Bush
21 Layshaft 3rd gear pinion
22 Layshaft 2nd gear pinion
23 Bush
24 Layshaft bottom gear pinion
25 Bush
26 Circlip
27 Driving dog
28 Quadrant

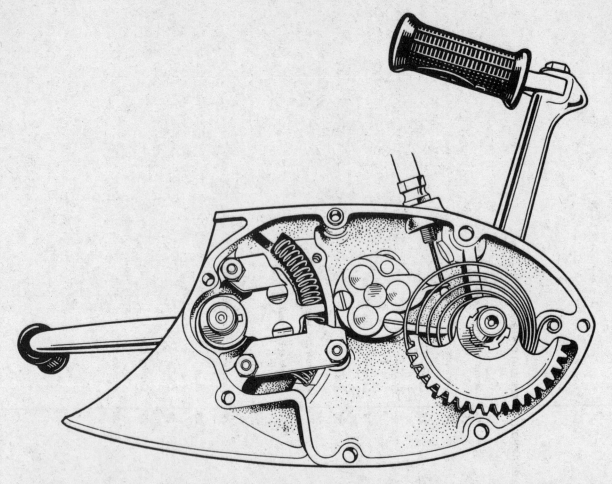

FIG.2.5. GEARBOX OUTER COVER, SHOWING GEARCHANGE MECHANISM, CLUTCH OPERATING MECHANISM AND KICKSTART QUADRANT

a tooth out, the gear change lever will lock in position during gear changing, necessitating a further strip down to rectify the alignment.

3 In the case of the five speed gearbox the cover should be arranged so that the second tooth of the quadrant, when it is in the raised position, is exactly in line with a centre line that passes through the footchange spindle housing.

4 Replace and tighten the two screws and bolt which retain the inner end cover to the gearbox shell.

5 Replace the oil pipe union over the stud near the base of the engine timing cover, using a new gasket and tighten the nut (and washer) holding the union in place. Refill the oil tank.

6 Temporarily replace the outer end cover and check that the gear change sequence is correct by operating the gear change lever and turning the rear wheel in unison. Any gear selection problems must be remedied at this stage by adjusting the position of the camplate quadrant in relation to that of the pre-set camplate.

7 When the gear change sequence is correct, remove the outer end cover and replace the kickstarter pinion and ratchet assembly, not forgetting the thin centre bush and the return spring. Place the tab washer and the securing nut on the end of the mainshaft and tighten the nut with a torque wrench to a setting of 45 ft lb whilst applying the back brake to prevent the mainshaft from turning. Bend the tab washer to lock the nut in position. Note that overtightening the nut may result in distortion or failure of the thin centre bush within the kickstarter pinion assembly. Re-insert the clutch pushrod and replace the right hand rear engine plate.

15 Gearbox reassembly - replacing the outer end cover and completing assembly

1 Coat the jointing surfaces of the outer and inner end covers with a thin layer of gasket cement and refit the end cover, after checking that the two locating dowels are in place. The kickstarter return spring must be tensioned one turn before the cover is placed in position, by rotating the kickstarter in an anticlockwise direction and holding it under tension until the cover is fully home. Note that it will be necessary to depress the kickstarter a further half turn whilst the cover is positioned so that the quadrant will clear the end stop. Replace the crosshead screws and the nuts to retain the cover in position.

2 Check that the kickstarter returns correctly and that all of the gears are selected in the correct sequence. It will be necessary to turn the rear wheel when making this latter check to ensure the gear pinions engage to their full depth.

3 Replace the cover plate in the rear portion of the primary chaincase, using a new paper gasket. Oil the seal in the centre of the cover plate and fit the assembly carefully so that the lip of the seal is not damaged Replace the primary drive by following the procedure given in Chapter 1, section 33 and adjust the clutch springs before the chaincase outer cover is replaced. Reconnect the clutch cable to the gearbox, engage the nipple with the clutch operating arm and adjust for a small amount of slack in the cable.

4 Refill the gearbox with 7/8 pint (500cc) SAE 50 oil (EP 90, 750 cc models) and the chaincase with 5/8 pint (350 cc) SAE 20 oil

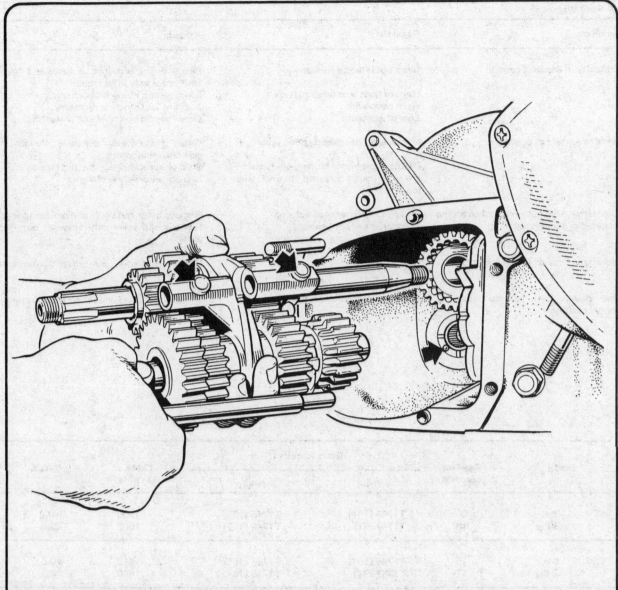

FIG.2.6. Reassembling the gearbox components: Arrows indicate complete rollers in position and thrust washers correctly located

5 Replace the gear change lever (if removed) and reconnect the speedometer drive cable if the machine is one of the earlier models. Replace the right hand footrest and the right hand exhaust system.

16 Changing speedometer drive gear combinations

Machines with engine numbers prior to DU 24875 only
1 Early machines have the speedometer drive taken from the right hand end of the gearbox layshaft and if any change is made in the overall gear ratios by changing the size of the gearbox final drive sprocket, as for sidecar work, the speedometer drive ratio has to be corrected in order to preserve accuracy.

2 The chart lists the part numbers of the speedometer drive gears required to maintain speedometer accuracy when the gearbox final drive sprocket is changed. It will be noted that only the final drive sprocket has to be changed when a sidecar is attached to the machine in order to achieve the optimum gear ratios.
3 The layshaft drive pinion is a push fit in the right hand end of the layshaft hollow and is secured with a pin that passes through the layshaft.
4 From engine number DU 24875 onwards, all machines were fitted with a rear wheel which contained provision for a hub-driven speedometer drive gearbox. This modification conveniently obviates the need to change the speedometer drive gears when the overall gear ratios of the machine are varied.

17 Fault diagnosis

Symptom	Reason/s	Remedy
Difficulty in engaging gears	Gears not indexed correctly	Check timing sequence of inner end cover (will occur only after rebuild).
	Worn or bent gear selector forks	Examine and renew if necessary.
	Worn camplate	Examine and renew as necessary.
	Low oil content	Check gearbox oil level and replenish.
Machine jumps out of gear	Mechanism not selecting positively	Check for sticking camplate plunger or gear change plungers.
	Sliding gear pinions binding on shafts	Strip gearbox and ease any high spots.
	Worn or badly rounded internal teeth in pinions	Replace all defective pinions.
Kickstarter does not return when engine is started or turned over	Broken kickstarter return spring	Remove outer end cover and replace spring.
	Kickstarter ratchet jamming	Remove end cover and renew all damaged parts.
Kickstarter slips on full engine load	Worn kickstarter ratchet	Remove end cover and renew all damaged parts.
Gear change lever fails to return to normal position	Broken or compressed return springs	Remove end cover and renew return springs.

Model		Gearbox Sprocket	Gears Required:		Cable R.P. Mile	Var. %
			Drive	Driven		
6T	Std.	20	T1744 (10T)	T1745 (15T)	1590	—0.6
	S/Car	18	T1747 (9T)	T1748 (15T)	1600	Zero
TR6	Std.	19	T1744 (10T)	T1745 (15T)	1610	+0.6
	S/Car	17	T1747 (9T)	T1748 (15T)	1640	+2.5
T120	Std.	19	T1744 (10T)	T1745 (15T)	1670	+4.2
	S/Car	17	T1747 (9T)	T1748 (15T)	1695	+5.9

TABLE OF SPEEDOMETER DRIVE GEAR COMBINATIONS

Note: The above chart only applies if the gearbox ratios, the number of teeth on the rear sprocket and the rear tyre size are as specified in "General Data" for the particular model, and % variation is calculated on the standard 1600 drive cable revolutions per mile.

Chapter 3 Clutch and Primary Transmission

Contents

Specifications

Clutch

Number of plates - inserted	6
plain	6
Pressure springs - number	3
free length	1 13/16 in
Bearing rollers - number	20
diameter	¼ in
length	0.231 in - 0.236 in
Pushrod - diameter of rod	7/32 in
length	11.822 in - 11.812 in

Sprockets (number of teeth)

Engine	29
Clutch	58

Primary chain

Size	3/8 in duplex endless **Triplex endless - 750 cc models**
Number of links	84

1 General description

The clutch is of the multi-plate type, designed to operate in oil. A synthetic friction material is used to line the inserted plates and a rubber cushion shock absorber is incorporated inside the centre drum of the clutch to even out transmission surges, particularly at low speeds. Drive from the engine sprocket is transmitted by a duplex or triplex roller chain. Because the gearbox is in unit with the engine, it is not possible to vary the centres between the engine mainshaft and the gearbox mainshaft. In consequence, a chain tensioner is incorporated so that the chain can be adjusted at regular intervals to take up wear.

2 Adjusting the clutch

1 The clutch can be adjusted by means of the handlebar lever cable adjuster, the cable adjuster in the top of the gearbox end cover, the pushrod adjuster in the centre of the clutch pressure plate and by varying the tension of the three clutch springs. In the latter case, the chaincase cover must be removed before adjustment can be effected.

2 To adjust the clutch, slacken off the handlebar lever adjuster so that there is an excess of free play in the lever, then unscrew the circular threaded plug in the outer face of the primary chaincase so that access can be gained to the adjuster in the centre of the pressure plate. Slacken the locknut and screw the adjuster inwards until the pressure plate just commences to lift. Back off the adjuster one complete turn, tighten the locknut and replace the threaded plug in the chaincase.

3 Adjust either the handlebar lever adjuster or the adjuster in the top of the gearbox end cover until there is approximately 1/8 in free play in the cable. This will ensure that there is no permanent loading on the clutch pushrod. Clutch adjustment should now be correct.

4 If the clutch still drags and the adjustment procedure described has produced no improvement, it will be necessary to remove the primary chaincase cover completely, in order to gain access to the three clutch spring adjusters in the pressure plate. This does not, however, apply to later models that have a removable plug in the centre of the chaincase clutch compartment. Before slackening the adjuster nuts, check that the drag is not caused by uneven tensioning of the pressure plate. This check is made by using the handlebar lever to slip the clutch and turning the clutch by depressing the kickstarter. Uneven tension will immediately be obvious by the characteristic 'wobble' of the pressure plate. The wobble can be corrected by retensioning the clutch springs individually until the pressure is even.

5 If the clutch still drags, the adjusters should be slackened off an equal amount at a time before a recheck is made. Do not slacken them off too much, or the clutch operation will become very light with the possibility of clutch slip when the engine is running.

6 Drag is often caused by wear of the clutch outer drum. The projecting tongues of the inserted plates will wear notches in the grooves of the drum, which will eventually trap the inserted

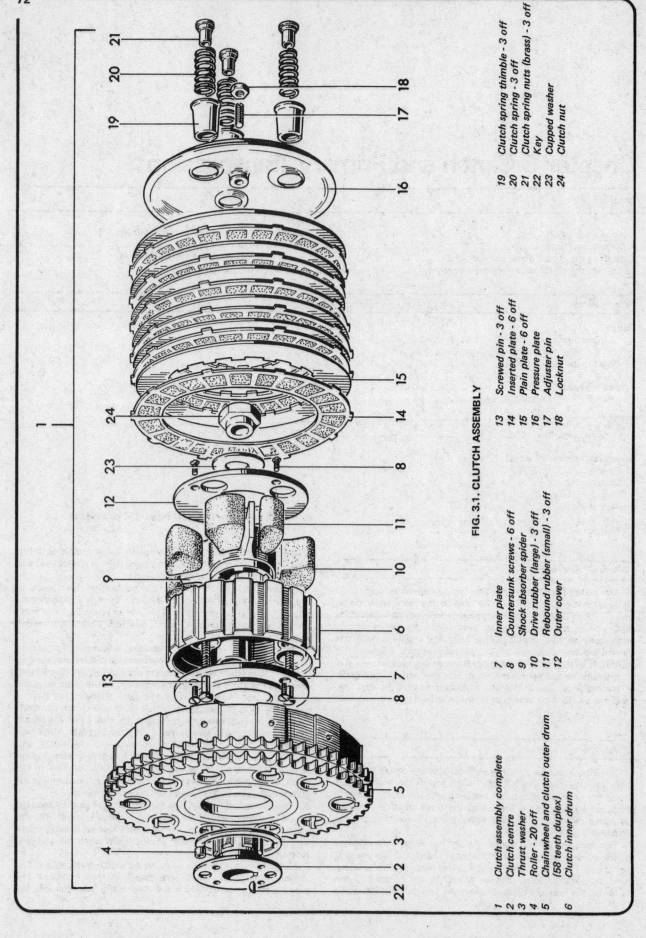

FIG. 3.1. CLUTCH ASSEMBLY

1 Clutch assembly complete
2 Clutch centre
3 Thrust washer
4 Roller - 20 off
5 Chainwheel and clutch outer drum (58 teeth duplex)
6 Clutch inner drum

7 Inner plate
8 Countersunk screws - 6 off
9 Shock absorber spider
10 Drive rubber (large) - 3 off
11 Rebound rubber (small) - 3 off
12 Outer cover

13 Screwed pin - 3 off
14 Inserted plate - 6 off
15 Plain plate - 6 off
16 Pressure plate
17 Adjuster pin
18 Locknut

19 Clutch spring thimble - 3 off
20 Clutch spring - 3 off
21 Clutch spring nuts (brass) - 3 off
22 Key
23 Cupped washer
24 Clutch nut

plates as the clutch is withdrawn and prevent them from separating fully. The accompanying photograph is an excellent example of such extended wear. In a case such as this, renewal of both the clutch chainwheel and the inserted clutch plates is necessary. If wear is detected in the early stages, it is possible to redress the grooves with a file until they are square once again, and to remove the burrs from the edges of the clutch plate tongues.

7 Heavy clutch operation is sometimes attributable to a cable badly in need of lubrication, or one in which the outer covering has become badly compressed through being trapped. Even a sharp bend in the cable will stiffen up the operation.

8 Clutch slip will occur when the clutch linings reach their limit of wear. If the reduction in lining thickness exceeds 0.030 in slip will occur and the inserted plates must be renewed. This necessitates dismantling the clutch, as described in Chapter 1, section 10. There is no necessity to dismantle any part of the generator or primary transmission assembly because the clutch plates can be withdrawn after the clutch pressure plate is removed.

9 It cannot be overstressed that a great many clutch problems are caused by failure to maintain the oil content of the chaincase at the correct level or by the use of a heavier grade of oil than that recommended by the manufacturers. It is also important that the chaincase oil is changed at regular intervals, to offset the effects of condensation.

2.6 Excessive wear of clutch drum by tongues of inserted plates

3 Examining the clutch plates and springs

1 When the clutch is dismantled for replacement of the clutch plates or during the course of a complete overhaul, this is an opportune time to examine all clutch components for signs of wear or damage.

2 As mentioned previously, the inserted clutch plates will have to be renewed when the reduction in thickness of the linings reaches 0.030 in. It is important to check the condition of the tongues at the edge of each plate which engage with the grooves in the clutch chainwheel. Even if the clutch linings have not approached the wear limit, it is advisable to renew the inserted plates if the width of the tongues is reduced as the result of wear.

3 Both plain and inserted plates should be prefectly flat. Check by laying them on a sheet of plate glass. Discoloration will occur if the plates have overheated; the surface finish of the plates is not too important provided it is smooth and the plates are not buckled.

4 The clutch springs will compress during service and take a permanent set. If any spring has reduced in length by more than 0.1 in, all three springs should be renewed - never renew one spring on its own. See Specifications section for the original free length measurement.

5 Check the hardened end of the clutch adjuster in the centre of the clutch pressure plate to ensure it has not softened from overheating or that the hardened surface is not chipped, cracked or worn away. This also applies to the ends of the clutch pushrod. Mysterious shortening of the pushrod, necessitating frequent clutch adjustment, can usually be attributed to incorrect adjustment of the clutch, such that the absence of free play in the clutch cable places a permanent load on the pushrod. This in turn causes the ends of the pushrod to overheat and soften, thus greatly accelerating the rate of wear.

6 The teeth of the clutch chainwheel should be examined since chipped, hooked or broken teeth will lead to very rapid chain wear. It is not possible to reclaim a worn chainwheel; the whole assembly must be renewed.

7 The clutch operating mechanism at the right hand side of the gearbox is unlikely to give trouble, or wear rapidly. The accompanying illustrations show the two types of mechanism fitted. Only the shape of the operating arm and the method of clutch cable attachment have been changed.

3.2 Check thickness of inserted plates when clutch is stripped

3.5 Hardened ends of pushrod must be in good condition

4 Examining the clutch shock absorber assembly

1 The shock absorber assembly is contained within the clutch inner drum and can be examined by unscrewing the three countersunk screws in the front cover plate. Remove the cover plate using a small screwdriver as a lever.

2 The shock absorber rubbers can be prised out of position commencing with the smaller rebound rubbers, to make the task easier. Avoid damage to the rubbers which may disintegrate in service if they are punctured or cracked. The centre 'spider' will be left in position and need not be disturbed unless it is cracked or broken. It is held in position by the nut that retains the clutch assembly to the gearbox mainshaft.

3 Reassemble the shock absorber assembly by reversing the dismantling procedure. Insert the large drive rubbers first and follow up with the smaller rebound rubbers. Fitting will be made easier if the smaller rubbers are smeared with household liquid detergent so that they can be slid into position. It is advisable to keep the rubbers free from oil, even though they are made of a synthetic material.

4 Before the cover plate screws are replaced, smear their threads with a sealant such as Loctite and tighten them fully. It is permissible to caulk the heads in position with a centre punch.

5 Engine sprocket - removal and replacement

1 Before the engine sprocket can be detached from the engine mainshaft, it will be necessary to remove the generator stator coils and rotor, and the complete clutch assembly because the engine sprocket and clutch chainwheel must be withdrawn together as a result of the endless chain used for the primary drive. Full details of the dismantling procedure are given in Chapter 1, section 10.

2 The engine sprocket is unlikely to require attention unless the teeth are chipped, hooked or broken - unlikely unless a chain breakage has caused the engine to lock up. Reference to Chapter 1, section 33 will show how to replace the primary drive after the sprocket has been renewed.

6 Adjusting the primary chain

1 The primary chain is of the duplex type and is endless because the engine and gearbox are built in unit with fixed centres. Provision for chain adjustment is made by incorporating a Weller type chain tensioner in the chaincase which is adjustable from a tunnel cast longitudinally in the base of the chaincase and sealed off by the chaincase drain plug.

2 To adjust the chain tensioners, first remove the filler cap from the top of the chaincase, to the rear of the cylinder barrel. Remove the chaincase drain plug, after placing a container below, to catch the oil. On some models it may be necessary to slacken the left hand footrest in order to gain better access.

3 Insert a screwdriver into the chaincase tunnel and turn it clockwise to increase the chain tension. Tension is correct when there is ½ inch free movement in the centre of the top chain run. Check with the sprockets in several different positions because a chain seldom wears evenly. The ½ inch play should be at the tightest point in the chain run.

4 Replace the chaincase drain plug, refill the chaincase with 5/8 pint, SAE 10W/30 (20W/50, 750 cc models) and replace the filler cap. Make sure the left hand footrest is tightened again if it has been slackened off.

5 If the chain tensioner reaches the end of its adjustment before the correct chain tension is achieved, the chain is due for renewal. It would be advisable to renew both the engine sprocket and the

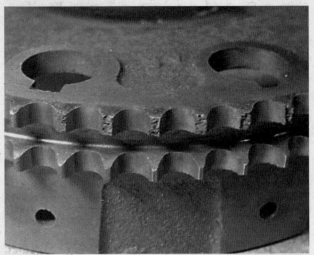

3.6 Broken teeth cause very rapid chain wear

clutch chainwheel at the same time whilst the primary transmission is dismantled completely. If old and new parts are run together, there is every possibility that the rate of wear will be accelerated.

6 The chain tensioner itself is unlikely to require attension unless the rubber facing becomes detached from the tensioner slipper. If this occurs, the tensioner should be renewed at the earliest opportunity. Damage of this nature is most likely to be caused if the chaincase is allowed to run low in oil content.

7 Removing and replacing the chaincase cover

1 The chaincase cover is secured to the chaincase casting by two domed nuts and eight crosshead screws. A paper gasket forms the seal between the two mating surfaces, both of which must be scrupulously clean if the chaincase is to remain oiltight

2 Before the chaincase cover can be removed, it is necessary to remove the drain plug and release the bolt securing the left hand silencer to the frame (if fitted) whilst the oil is draining off, so that the exhaust system clears the cover. Remove the left hand footrest. There is no necessity to remove the rear brake pedal because this can be depressed whilst the cover is lifted clear.

3 Check that the cover is not distorted before coating the jointing surface with gasket cement. If the distortion is only slight, the chaincase cover can be rubbed down until it is flat. Coat the jointing face of the chaincase casting with which the cover mates.

4 Position the paper gasket on the cover and replace all screws and nuts. If the screw heads are damaged, it is worth renewing them with Allen screws of the same thread and length. These are often available from Triumph dealers as a replacement kit. When all screws and nuts are located correctly and fingertight, final tightening can be completed. Refit the drain plug.

5 If the cover still leaks after the oil content (5/8 pint SAE 10W/30 (20W/50 750 cc models) has been added, check that it is not leaking from the threads of the lower of the two domed nuts and running down the casting to give the impression of a leaking gasket. This form of leakage can be cured by placing a fibre washer under each nut before tightening.

6 Replace the filler plug and check that the circular plug in the centre of the chaincase is tight.

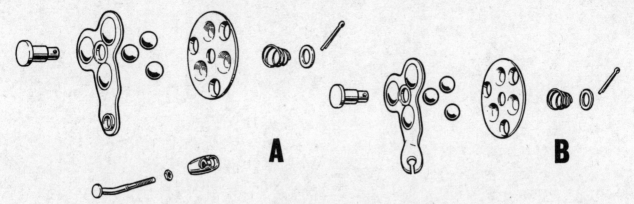

FIG.3.2. (A) Clutch operating mechanism up to DU66246   (B) From DU66246 onward

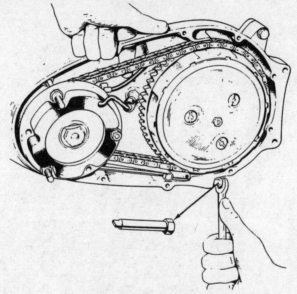

FIG.3.3. Adjusting chain tension

8 Fault diagnosis

Symptom	Reason/s	Remedy
Engine speed increases but machine does not respond	Clutch slip	Check clutch adjustment. If correct, suspect worn linings and/or weak springs.
Difficulty in engaging gears. Gear changes jerky and machine creeps forward, even when clutch is withdrawn fully	Clutch drag Clutch plates worn and/of clutch drums Clutch assembly loose on mainshaft	Check adjustment for too much play. Check for burrs on clutch plate tongues and indentations in clutch drum grooves. Check tightness of retaining nut. If loose, fit new tab washer and retighten.
Operating action stiff	Damaged, trapped or frayed control cable. Cable bends too acute Pushrod bent Spring adjusters too tight	Check cable and replace if necessary. Re-route cable to avoid sharp bends. Replace. Slacken adjusters and check clutch does not slip.
Clutch needs frequent adjustment	Rapid wear of pushrod	Leave slack in cable to prevent continual load on pushrod. Renew rod because over-heating has softened ends.
Harsh transmission	Worn chain and/or sprockets	Replace.
Transmission surges at low speeds	Worn or damaged shock absorber rubbers	Dismantle clutch shock absorber and renew rubbers.

Chapter 4 Fuel system and lubrication

Contents

Specifications

Carburettor	Models: (to 1968)	T120	T120TT *	TR6
Amal type			Monobloc	
Number		389/203	389/95	389/239
Bore		$1^{1}/8$ in	$1^{3}/16$ in	$1^{3}/16$ in
Main jet		260	230	230
Pilot jet		25	25	25
Needle jet		0.106	0.106	0.106
Needle type		D	D	D
Needle position		3	2	1
Throttle valve		2½	2	4
Air cleaner type		none	felt or cloth	felt

* Also applicable to TR6C, TR6R and T120R models

	Models: (1968 on)	T120	T120R & T120RV	TR6 & TR6R	TR7V	T140V
Amal type				Concentric		
Number		R930/9 and L930/10	R930/66R and L930/67	930/23	R930/89	R930/92 and L930/93
Bore (mm)		30	30	30	30	30
Main jet		220	180	230	260	190
Needle jet		0.106	0.106	0.106	0.106	1.106
Needle type		STD	STD	STD	STD	STD
Needle position		2	1	2	1	1
Throttle valve		2½	3	3½	3½	3
Air cleaner type		none	none	cloth	cloth	cloth

Fuel tank capacities	4 gallons	3½ gallons * or 2½ gallons +	4 gallons

* TR6R models   + TR6C, T120R, T120TT and TR6R models (alternatives)

Fuel

Octane rating (minimum) ... ...   97 Premium grade (BS 4040 - 4 Star)

Oil tank capacity:

Oil tank models ... ...   5 Imp pints (2.84 litre)
Oil in frame models ... ...   4 Imp pints (2.27 litre)

Oil pressure readings

Normal running	65/80 psi	all models
Idling	20/25 psi	all models

Oil pressure release valve

Plunger diameter	0.5605 in - 0.5610 in	all models
Working clearance	0.001 in - 0.002 in	all models
Operating pressure	60 psi	all models
Spring length (free)	1 3/8 in	all models

Oil pump

Ball valve spring length	½ in	all models
Ball diameter	7/32 in	all models

1 General description

The fuel system comprises a petrol tank from which petrol is fed by gravity to the float chamber(s) of the carburettor(s). Two petrol taps, with built-in gauze filter, are located one each side beneath the rear end of the petrol tank. For normal running the right hand tap alone should be opened except under high speed and racing conditions. The left hand tap is used to provide a reserve supply, when the main contents of the petrol tank are exhausted.

For cold starting the carburettor(s) incorporate an air slide which acts as a choke controlled from a lever on the handlebars. As soon as the engine has started, the choke can be opened gradually until the engine will accept full air under normal running conditions.

Lubrication is effected by the 'dry sump' principle in which oil from the separate oil tank is delivered by gravity to the mechanical oil pump located within the timing chest. Oil is distributed under pressure from the oil pump through drillings in the crankshaft to the big ends where the oil escapes and is fed by splash to the cylinder walls, ball journal main bearings and the other internal engine parts. Pressure is controlled by a pressure release valve, also within the timing chest. After lubricating the various engine components, the oil falls back into the crankcase, where it is returned to the oil tank by means of the scavenge pump. A bleed-off from the return feed to the oil tank is arranged to lubricate the rocker arms and valve gear, after which it falls by gravity via the pushrod tubes and the tappet blocks, to the crankcase. An additional, positive oil feed is arranged from drillings in the timing cover to lubricate the exhaust tappets. It will be noted that the oil pump is designed so that the scavenge plunger has a greater capacity than the feed plunger, this is necessary to ensure that the crankcase is not flooded with oil, and that any oil drain-back whilst the machine is standing is cleared quickly, immediately the engine starts.

2 Petrol tank - removal and replacement

1　The petrol tank is secured to the frame by two studs underneath the nose, one on each side. These studs project through two short brackets welded to the frame and are cushioned by rubber washers to damp out vibration. The tank is retained at the front by two self-locking nuts and washers which thread onto the studs. Early models have two bolts, threaded directly into the tank and wire-locked together. The rear mounting takes the form of a lug welded to the rear of the tank which matches with a threaded hole in the top portion of the frame, close to the nose of the dual seat. Anchorage is provided by a bolt which passes through shaped rubbers to provide a flexible mounting.

2　When the bolt and two nuts are removed and the two fuel pipe unions disconnected at their joint with the petrol taps, the tank can be lifted from the machine. Make sure the shaped rubbers are not lost, since they will be displaced as the tank is removed.

3　When replacing the tank, special care must be taken to ensure none of the carburettor control cables are trapped or bent to a sharp radius. Apart from making control operation much heavier,

there is risk that the throttle may stick since there is minimum clearance between the underside of the petrol tank and the top frame tube.

4　Models exported to the USA have reflectors fitted below the front of the petrol tank secured by the front mounting nuts or bolts. The reflector units must be removed first. If necessary the reflector lenses can be prised out of position to gain better access to the nuts or bolts.

3 Petrol taps - removal and replacement

1　The petrol taps are threaded into inserts in the rear of the petrol tank, at the underside. Neither tap contains provision for turning on a reserve quantity of fuel. It is customary to use the right hand tap only so that the left hand tap will supply the reserve quantity of fuel, unless the machine is used for high speed work or racing. In these latter cases, it is essential to use both taps in order to obviate the risk of fuel starvation.

2　Before either tap can be unscrewed and removed, the petrol tank must be drained. When the taps are removed each gauze filter, which is an integral part of the tap body, will be exposed.

3　When the taps are replaced, each should have a new sealing washer to prevent leakage from the threaded insert in the bottom of the tank. Do not overtighten; it should be sufficient just to commence compressing the fibre sealing washer.

4 Petrol feed pipes - examination

1　Plastic feed pipes of the transparent variety are used with a union connection to each petrol tap and a push-on fit at the carburettor float chamber.

2　After lengthy service, the pipes will discolour and harden gradually due to the action of the petrol. There is no necessity to renew the pipes at this stage unless cracks become apparent or the pipe becomes rigid and 'brittle'.

5 Carburettor(s) - removal

1　Both single and twin carburettor fitments have been used depending on the version. Early models used the Amal Monobloc carburettor(s) whilst later and now current versions use the Amal Concentric carburettor(s). Both types are described here but special emphasis is given to the concentric because it is, by now, the most usual fitment or replacement.

2　Before removing a carburettor it is first necessary to detach the mixing chamber top which is retained by two small screws and lift away the top complete with the control cables, throttle valve and air slide assemblies. The petrol pipe can then be pulled off the push connection at the float chamber (or the union complete detached) and, after detaching the two retaining nuts and shakeproof washers, the complete carburettor, may be removed from the cylinder head.

6 Carburettor(s) - dismantling, examination and reassembly

Amal concentric carburettor only

1　To remove the float chamber, unscrew the two crosshead screws on the underside of the mixing chamber. The float

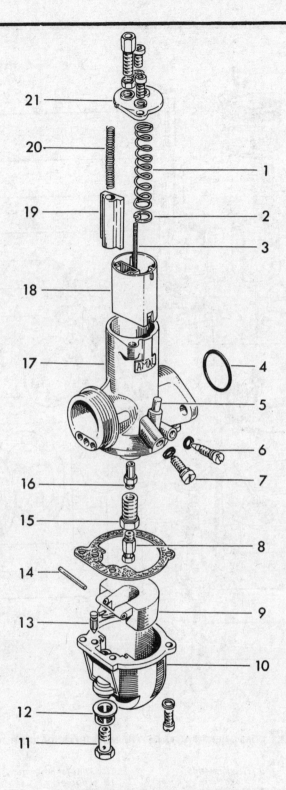

**FIG.4.1. COMPONENT PARTS OF THE CONCENTRIC
CARBURETTOR**

1	Throttle return spring	7	Throttle stop screw	13	Float needle	19 Air slide (choke)
2	Needle clip	8	Main jet	14	Float hinge	20 Air slide return spring
3	Needle	9	Float	15	Jet holder	21 Mixing chamber top
4	'O' ring	10	Float chamber	16	Needle jet	
5	Tickler	11	Banjo union bolt	17	Mixing chamber body	
6	Pilot jet screw	12	Filter	18	Throttle valve (slide)	

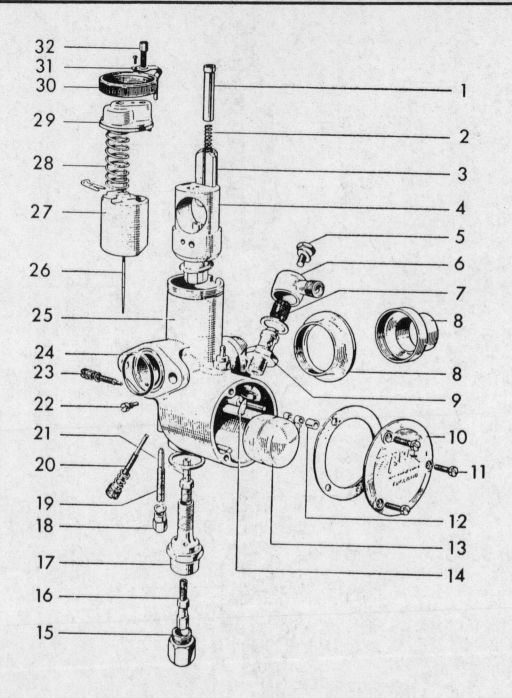

FIG.4.2. COMPONENT PARTS OF THE MONOBLOC CARBURETTOR

1	Air valve guide	9	Needle setting	18	Pilot jet cover nut	27	Throttle slide
2	Air valve spring	10	Float chamber cover	19	Pilot jet	28	Throttle spring
3	Air valve	11	Cover screw	20	Throttle stop screw	29	Top
4	Jet block	12	Float spindle bush	21	Needle jet	30	Cap
5	Banjo bolt	13	Float	22	Locating peg	31	Click spring
6	Banjo	14	Float needle	23	Pilot air screw	32	Adjuster
7	Filter gauze	15	Main jet cover	24	'O' ring seal		
8	Air filter connection (top) or air intake tube	16	Main jet	25	Mixing chamber		
		17	Main jet holder	26	Jet needle		

chamber can then be pulled away complete with float assembly and sealing gasket. Remove the gasket and lift out the horseshoe-shaped float, float needle and spindle on which the float pivots.

2   When the float chamber has been removed, access is available to the main jet, jet holder and needle jet. The main jet threads into the jet holder and should be removed first, from the underside of the mixing chamber. Next unscrew the jet holder which contains the needle jet. The needle jet cannot be removed until the jet holder has been unscrewed and removed from the mixing chamber because it threads into the jet holder from the top. There is no necessity to remove the throttle stop or air adjusting screws.

3   Check the float needle for wear which will be evident in the form of a ridge worn close to the point. Renew the needle if there is any doubt about its condition, otherwise persistent carburettor flooding may occur.

4   The float itself is unlikely to give trouble unless it is punctured and admits petrol. This type of failure will be self-evident and will necessitate renewal of the float.

5   The pivot needle must be straight - check by rolling the needle on a sheet of plate glass.

6   It is important that the gasket between the float chamber and the mixing chamber is in good condition if a petrol tight joint is to be made. If it proves necessary to make a replacement gasket, it must follow the exact shape of the original. A portion of the gasket helps retain the float pivot in its correct location; if the pin rides free it may become displaced and allow the float to rise, causing continual flooding and difficulty in tracing the cause. Use Amal replacements whenever possible.

7   Remove the union at the base of the float chamber and check that the inner nylon filter is clean. All sealing washers must be in good condition.

8   Make sure that the float chamber is clean before replacing the float and float needle assembly. The float needle must engage correctly with the lip formed on the float pivot; it has a groove that must engage with the lip. Check that the sealing gasket is placed OVER the float pivot spindle and the spindle is positioned correctly in its seating.

9   Check that the main jet and needle jet are clean and unobstructed before replacing them in the mixing chamber body. Never use wire or any pointed instrument to clear a blocked jet, otherwise there is risk of enlarging the orifice and changing the carburation. Compressed air provides the best means, using a tyre pump if necessary.

10   Before refitting the float chamber, check that the jet holder and main jet are tight. Do not invert the float chamber, otherwise the inner components will be displaced as the retaining screws are fitted. Each screw should have a spring washer to obviate the risk of slackening.

11   When replacing the carburettor, check that the O ring seal in the flange mounting is in good condition. It provides an airtight seal between the carburettor flange and the cylinder head flange to ensure the mixture strength is constant. Do not overtighten the carburettor retaining nuts for it is only too easy to bow the flange and give rise to air leaks. A bowed flange can be corrected by removing the O ring and rubbing down on a sheet of fine emery cloth wrapped around a sheet of plate glass, using a circular motion. A straight edge will show if the flange is level again, when the O ring can be replaced and the carburettor refitted.

12   Before the mixing chamber top is replaced, check the throttle valve for wear. A worn valve is often responsible for a clicking noise when the throttle is opened and closed. Check that the needle is not bent and that it is held firmly by the clip.

Amal monobloc carburettor only

Early models were fitted with the Amal monobloc carburettor which preceded the concentric type, currently in use. Since the two designs of carburettor differ in a number of respects, revised procedure is necessary when dismantling, examining and reassembling the monobloc instrument.

13   The float chamber is an integral part of the monobloc carburettor and cannot be separated. Access is gained by removing three countersunk screws in the side of the float

2.2   When replacing the petrol tank, check no cables are trapped

5.2a   To release needle and slide assemblies, detach carburettor top

5.2b   On 'Concentric' carburettor, float chamber is part of mixing chamber

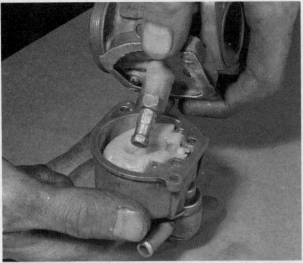

6.1a Float chamber is released by removing two retaining screws on underside

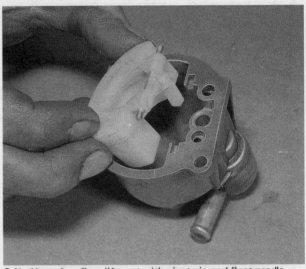

6.1b Horseshoe float lifts out with pivot pin and float needle

6.2 Main jet threads into jet holder

6.7 Banjo union at float chamber base contains nylon filter

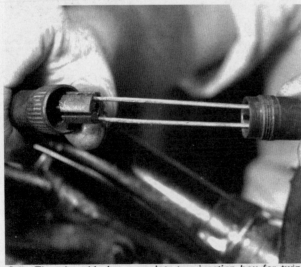

8.1 Throttle cable has a one-into-two junction box for twin carburettors

11.2 Pressure release plunger must be clean and free in housing

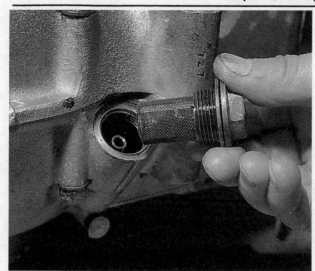

12.2 Crankcase filter is of gauze mesh type

chamber and removing the end cover and gasket. Remove the small brass distance piece and the float needle which will free from its seating as the float is withdrawn.

14 The main jet threads into the main jet holder which itself is screwed into the main body of the mixing chamber. Removal of the lower main jet cover gives access to the main jet. If the hexagonal nut above the jet cover is unscrewed, the main jet holder can be detached and the needle jet unscrewed from the upper end. The pilot jet has its own separate cover nut. When removed, the jet can be unscrewed. It is threaded at one end and has a screwdriver slot.

15 Unless internal blockages are suspected, or the body is worn badly, there is no necessity to remove the jet block, which is a tight push fit within the mixing chamber body. It is removed by pressing upward, through the orifice of the main jet holder, after removing the small locating peg which threads into the carburettor body. Extreme care must be exercised to prevent distorting either the jet block or the carburettor body which is cast in a zinc-based alloy.

16 Check the float needle for wear by examining it closely. If a ridge has worn around the needle, close to the point, the needle should be discarded and a new one fitted.

17 The float is unlikely to give trouble unless it is punctured, in which case a replacement is essential. Do not omit to fit the small brass distance piece on the float pivot, after the float has been inserted. If this part is lost, there is nothing to prevent the float moving across to the float chamber end cover and binding - a fault that will give rise to intermittent flooding and prove difficult to pinpoint.

18 There must be a good seal between the float chamber end cover and the float chamber. Always use a new gasket when the seal is broken to obviate the risk of continual petrol leakage.

19 Do not omit to inspect and, if necessary, clean the nylon filter within the float chamber union. When replacing the filter, position it so that the gauze is facing the inflow of petrol. On some of the earlier filters, the plastic dividing strips between the gauze segments are somewhat wide and could impede the flow of petrol under full flow conditions.

20 As stressed in the preceding part of this section, do not use wire or any pointed object to clear blocked jets. Compressed air should be used to clear blockages; even a tyre pump can be utilised if a compressed air line is not available.

21 The monobloc carburettor has an O ring in the centre of the mounting flange which must be in good condition if air leaks are to be excluded. If the flange is bowed, as the result of previous overtightening, the O ring should be removed and the flange rubbed down on fine emery cloth wrapped around a sheet of plate glass. Rub with a rotary motion and when a straight edge

shows the flange is level again, the O ring can be replaced.

7 Carburettor(s) - checking the settings

1 The various sizes of jets and that of the throttle slide, needle and needle jet are predetermined by the manufacturer and should not require modification. Check with the Specifications list if there is any doubt about the values fitted.

2 Slow running is controlled by a combination of the throttle stop and air regulating screw settings. Commence by screwing the throttle stop screw(s) inward so that the engine runs at a fast tickover speed. Adjust the pilot air screw setting(s) until the tickover is even, without either misfiring or 'hunting'. Screw the throttle stop screw(s) outward again until the desired tickover speed is obtained, then recheck with the pilot air screw(s) so that the tickover is as even as possible. Always make these adjustments with the engine at normal running temperature and remember that an engine fitted with high-lift cams is unlikely to run evenly at very low speeds no matter how carefully the adjustments are made.

3 If desired, there is no reason why the throttle stop screw(s) should not be lowered so that the engine will stop completely when the throttle is closed. Some riders prefer this arrangement so that the maximum braking effect of the engine can be utilised on the over-run.

4 As an approximate guide, up to 1/8 throttle is controlled by the pilot jet, from 1/8 to 1/4 throttle by the throttle valve cutaway, from 1/4 to 3/4 throttle by the needle position and from 3/4 to full throttle by the size of the main jet. These are only approximate divisions; there is a certain amount of overlap.

8 Balancing twin carburettors

1 Twin carburettors are fitted to the 650 cc and 750 cc Bonneville models, using left and right handed carburettors. There is a balance pipe linking both carburettor inlet ports to improve tickover. A one-into-two throttle cable and air slide cable assemblies are used, each with its own junction box. The junction box components are made of a plastic material; no maintenance is necessary.

2 Before commencing the balancing operation, it is essential to check that both carburettors operate simultaneously. Place a finger inside the bell mouth of each carburettor intake in turn and check when the throttle valve commences to move as the twist grip is rotated. Both slides should begin to rise at exactly the same time; if they do not, use the cable adjusters to ensure the moment of lift coincides. It is important that the throttle stop screws are slackened off during this operation to obviate the risk of a false reading.

3 Cross-check by noting the points at which the throttle slides lift completely and again, adjusting if necessary.

4 Start the engine and when it is at running temperature, stop it and remove one spark plug lead. Restart the engine and adjust the air regulating screw and throttle stop screw of the OPPOSITE cylinder as detailed in Section 7.2 until the desired tickover speed is obtained. Stop the engine again, replace the spark plug lead and repeat the whole operation with the other cylinder and carburettor.

5 When both spark plug leads are replaced, it is probable that the tickover speed will be too high. It can be reduced to the desired level by unscrewing both throttle stop screws an identical amount and rechecking to ensure both throttle valves still lift simultaneously.

9 Air cleaner - removal and replacement

1 Some models are fitted with an air cleaner which takes the form of a separate, circular housing attached direct to the carburettor air intake, or an oblong box with rounded corners, mounted across the frame. Two types of filter element have been

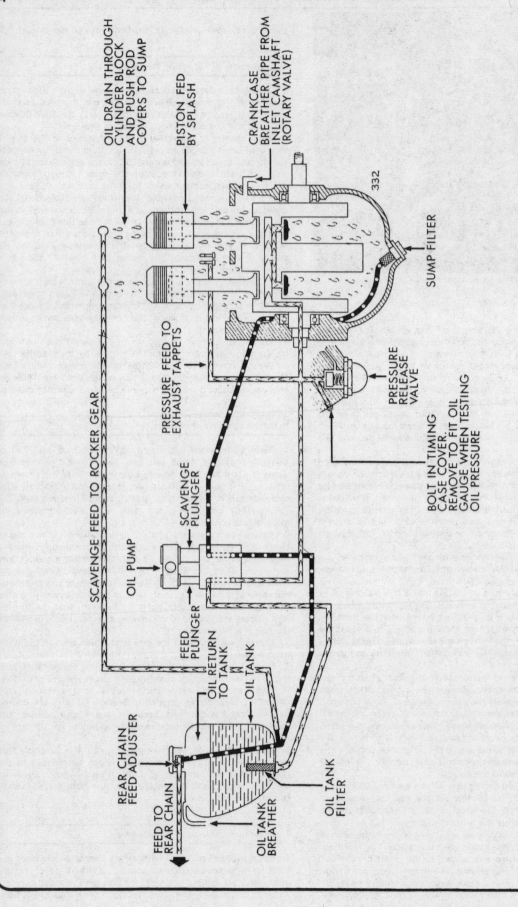

OIL DRAIN THROUGH CYLINDER BLOCK AND PUSH ROD COVERS TO SUMP

PISTON FED BY SPLASH

CRANKCASE BREATHER PIPE FROM INLET CAMSHAFT (ROTARY VALVE)

332

SUMP FILTER

SCAVENGE FEED TO ROCKER GEAR

PRESSURE FEED TO EXHAUST TAPPETS

PRESSURE RELEASE VALVE

BOLT IN TIMING CASE COVER. REMOVE TO FIT OIL GAUGE WHEN TESTING OIL PRESSURE

OIL PUMP

SCAVENGE PLUNGER

FEED PLUNGER

OIL RETURN TO TANK

OIL TANK

REAR CHAIN FEED ADJUSTER

FEED TO REAR CHAIN

OIL TANK BREATHER

OIL TANK FILTER

FIG.4.3. ENGINE LUBRICATION

employed, a convoluted paper element or one formed from cloth or felt.

2 None of the filter elements should be soaked in oil. It is sufficient to detach the paper element and blow it clean with compressed air or in the case of the cloth or felt elements, to wash them in paraffin and allow them to drain thoroughly before replacing.

3 On no account run the machine with the air cleaner disconnected unless the carburettor has been re-jetted to suit. When an air cleaner is fitted, it is customary to reduce the size of the carburettor main jet, in order to compensate for the enrichening effect of the air cleaner element. In consequence, a permanently-weakened mixture will result if the air cleaner is detached; this will cause failure of the valves and/or piston crown.

10 Exhaust system - general

1 Three separate types of exhaust system are fitted to the 650/750 cc twins, the type depending on the specification of the model concerned. Most machines have a downswept exhaust system of the two pipe and two silencer type, which may or may not be joined together by a balance pipe close to the exhaust ports. The export Trophy models have twin pipes and silencers upswept on the left hand side of the machine. The export Bonneville TT Special was supplied with twin downswept pipes that terminate close to the rear tyre. No silencers were supplied with this model, which was intended for track racing.

2 It is important that the exhaust pipes are a good fit over the stubs which project from the cylinder head. An air leak at this joint will cause the engine to backfire on the over-run.

3 If renewal of either the silencers or exhaust pipes should prove necessary, fit genuine replacement parts of Triumph origin. Although there are many alternative exhaust systems and silencers on the market, it does not necessarily follow that they will improve, let alone maintain, the already high standard of performance. A changed exhaust note is not always indicative of greater speed; in a great many cases the fitting of an exhaust system, not compatible with the engine characteristics, will result in a marked drop in performance.

4 Current models are fitted with an entirely new type of silencer, evolved to give a significant reduction in noise level without undue power loss. These silencers are recognisable by their long, tapered shape and reverse cone end. They can be fitted as direct replacements to earlier versions.

11 Engine lubrication - removing and replacing the oil pressure release valve

1 Oil pressure is controlled by a pressure release valve located within the timing chest on the right hand side of the engine. Oil pressure, if suspect, can be verified by attaching an oil gauge to the blanking off plug in the forward-facing edge of the timing cover, using an adaptor. From a cold start, the pressure may rise as high as 80 psi initially but when the engine is at normal running temperature, this reading should drop to 20 or 25 psi at tickover speeds. Normal running pressure should be within the 65 to 80 psi range.

2 If satisfactory readings are not obtained, the engine should be stopped and the pressure release valve dismantled. It is preferable to remove the complete unit by unscrewing the hexagonal nut closest to the timing chest and then dismantle the unit whilst it is clamped in a vice. When the domed cap is removed in this fashion, it will release the spring and the plunger in which the spring seats.

3 Wash the dismantled components in a petrol/paraffin mix and ensure that they are dry before examination. Check that the gauze filter in the end of the unit is not blocked or damaged and that the plunger is free from score marks. The main body of the unit should be unscored internally.

4 Check the free length of the plunger spring. If the length is

less than $1^{17/32}$ in the spring must be renewed.

5 Reassemble the unit using new fibre sealing washers. Replace the unit in the timing chest and tighten it fully, with a new fibre washer under the hexagon nut. Start the engine and if the oil pressure readings are still low, give attention to the oil pump itself and the filters in the oil tank and the crankcase.

6 Engines prior to DU 13375 employ the older type of pressure release valve in which an indicator button extends from the dome nut. This is dismantled in similar fashion; additional parts to check for wear or damage are the rubber sleeve over the indicator button shaft and the tiny O ring seal near the end of the indicator button. Note that in this design of release valve, the plunger spring is in two parts.

12 Engine lubrication - location and examination of oil filters

1 Two filters are included in the engine lubrication system, both of the gauze mesh type. One is located in the underside of the crankcase at an angle, blanked off by a cap and guide rod accessible from the left hand side of the machine. The other takes the form of an extension of the oil outlet (feed) pipe from the oil tank, within the body of the tank itself. On the post 1970 models, the oil is contained within the frame.

2 The crankcase filter is withdrawn with the blanking off cap when the latter is unscrewed from the underside of the crankcase. It should be washed with a petrol/paraffin mix and replaced, making sure the cap sealing washer is in good condition.

3 The filter within the oil tank or frame is most unlikely to require attention unless some heavily contaminated oil has been added by accident. To remove the filter, unscrew the hexagon nut above the main feed union. The filter takes the form of a metal gauze, which should be cleansed with petrol and permitted to dry. If the filter is blocked, it will immediately be evident when the main oil feel pipe is detached before the tank is drained.

13 Engine lubrication - removal, examination and replacement of oil pump

1 The oil pump is located within the timing cover, which must be removed to gain access. See Chapter 1, section 8. The oil pump is retained by two conical nuts which should not be removed until the oil tank is drained. When both nuts are removed, the oil pump is free to be drawn off the mounting studs.

2 The part subject to most wear is the drive block slider, which should be renewed to maintain pump efficiency. Wear will be obvious immediately on examining the block.

3 The oil pump plungers are not subject to wear because they are continuously immersed in oil. They should, however, be checked for score marks and any undue slackness in the bores. The normal running clearance is up to 0.0005 inch in the case of both the scavenge and feed plungers.

4 Remove the square-headed plugs from the bottom of each plunger housing and check that in each case the ball valve is not sticking to its seat. The springs should have a free length of ½ inch; if they have taken a permanent set they must be renewed.

5 When reassembling the pump, prime both plunger bores with engine oil and check that oil is forced through the outlet ports as the plungers are inserted and depressed. Check that the oil levels within the bores do not fall as the plungers are raised. If they do fall, this denotes a badly seating ball valve at the base of the plunger. Dismantle the ball valve assembly again and clean the seating. The seating can be restored by tapping the ball on its seating with a punch, but only if the pump body is made of brass. Replace the spring and end cap, then recheck again.

6 Always fit a new gasket when the pump is refitted to the timing chest and check that the oil holes align correctly. Do not use gasket cement of any kind, otherwise there is danger of restricting or even blocking the oilways. Check that the drive block slider engages correctly with the peg on the inlet camshaft pinion, before replacing and tightening the conical

retaining nuts.

7 Clean off and use new gasket cement on the timing cover when it is replaced to ensure an oiltight joint. Do not forget to replace the blanking off plug when the oil pressure gauge is removed.

8 Remember that most lubrication troubles are caused by failure to change the engine oil at the prescribed intervals. Oil is cheaper than bearings!

14 Fault diagnosis

Symptom	Reason/s	Remedy
Excessive fuel consumption	Air filter choked, damp or oily	Check and if necessary renew.
	Fuel leaking from carburettor	Check all unions and gaskets.
	Float needle sticking	Float needle seat needs cleaning.
	Worn carburettor	Renew.
Idling speed too high	Throttle stop screw in too far	Re-adjust screw.
	Carburettor top loose	Tighten top.
Engine does not respond to throttle	Mixture too rich	Check for displaced or punctured float.
Engine dies after running for a short while	Blocked air vent in filler cap	Clean.
	Dirt or water in carburettor	Remove and clean float chamber.
General lack of performance	Weak mixture; float needle stuck in seat	Remove float chamber and check.
	Leak between carburettor and cylinder head	Bowed flange; rub down until flat and replace O ring seal.
	Fuel starvation	Turn on both petrol taps for fast road work.

Chapter 5 Ignition system

Contents

Specifications

Ignition coils

Make	Lucas
Type	MA12
Voltage	12 volt

Note: Machines fitted with a six volt AC alternator require special coils

Contact breakers

Make	Lucas
Type	4CA or 6CA (6CA after engine number DU 66246)
Gap	0.014 - 0.016 in (0.35 - 0.40 mm)

Spark plugs

Make and type	Champion N3	KLG FE100	LODGE 2HLN	NGK B8ES
Size	14 mm	14 mm	14 mm	14 mm
Reach	¾ in	¾ in	¾ in	¾ in

Gap:
Engine nos. DU101 - DU66245 (1963-67 models fitted with 4CA contact breaker assembly)	0.020 in (0.50 mm)
Engine no. DU66246 onwards (all models fitted with 6CA or 10CA contact breaker assembly)	0.025 in (0.64 mm)

Ignition timing

Static	14° BTDC
Fully advanced	38° BTDC
* Static	29° BTDC - engine numbers prior to DU 66246
* Fully advanced	39° BTDC - engine numbers prior to DU 66246

* AC magneto ignition equipment. Applies also to USA only export models TR6R, TR6C, T120R and T120TT (up to DU 66245)

1 General description

The spark necessary to ignite the petrol/air mixture in each combustion chamber is derived from a battery and coil, used in conjunction with a contact breaker to determine the precise moment at which the spark will occur. As the points separate the circuit is broken and a high tension voltage is developed across the points of the spark plug which jumps the air gap and ignites the mixture. Each cylinder has its own ignition circuit, hence the need for two separate ignition coils and a twin contact breaker assembly.

When the engine is running, the surplus current produced by the generator is converted into direct current by the rectifier and used to charge the battery. The six volt system used for the earlier models contains provision for emergency starting if the battery is fully discharged. This facility is not required in the case of the twelve volt system because the generator provides sufficient current for the initial start under similar circumstances.

Generator output does not correspond directly to engine rpm and is regulated by a diode in circuit, eliminating the need for an electro-mechanical device such as a voltage regulator. All coils in the system are brought into operation only if there is a heavy electrical load when all lights are in use.

Some models, such as the TR6SC and the T120TT Special, have an AC ignition system that does not require a battery in the circuit. These machines, all with engine numbers prior to DU 24875, have no lighting equipment fitted.

2 Checking generator output

Specialised test equipment of the multi-meter type is essential to check generator output with any accuracy. It is unlikely that the average owner will have access to this type of equipment or instruction in its use. In consequence, if generator performance is suspect, it should be checked by a Triumph agent or an auto-electrical expert.

3 Ignition coils - checking

1 Ignition coil is a sealed unit, designed to give long service without need of attention. A twin coil system is used on the Triumph unit-construction twins, with the coils mounted one on each side of a bracket that joins the upper and lower frame tubes. It is necessary to remove the petrol tank in order to gain access. This position was later changed to under the seat on oil-in-the-frame models.

2 To check whether a coil is defective, disconnect the spark plug lead from the plug concerned and turn the engine over until the contact breaker points that relate to the coil being tested are closed (check with colour coding of wire). Switch on the ignition and hold the plug lead about 3/16 in away from the cylinder head. If the coil is in good order, a strong spark should jump the air gap between the end of the plug lead and the cylinder head, each time the points are flicked open.

3 A coil is most likely to fail if the outer casing is compressed or damaged in any way. Fine gauge wire is used for the secondary winding and this will break easily if subjected to any strain by a damaged casing.

4 Contact breakers - adjustment

1 The contact breaker points are located behind the chromium plated cover attached to the forward end of the timing cover by two crosshead screws. Remove both screws and withdraw the cover.

2 Two types of contact breaker assembly have been used. Machines with engine numbers prior to DU 66246 have the Lucas 4CA type containing condensers. Later models utilise the Lucas 6CA assembly, which is more accessible because the condensers have been transferred to another location.

3 In each case, the correct contact breaker gap is 0.015 in with the points open fully. To adjust the gap on the 4CA unit, slacken the slotted nut that secures the stationary contact point and move the contact either inwards or outwards until the gap is correct. Tighten the nut and recheck that the setting is still correct. Repeat this procedure for the other set of points.

4 The 6CA unit has the fixed contact point secured by a locking screw which must be slackened first. Adjustment is effected by turning an eccentric screw in the forked end of each contact breaker point assembly, before tightening the locking screw. Recheck that the setting is correct before repeating the procedure for the second set of points. With the 6CA unit, checking whether the points are open fully is simplified by aligning a scribe mark on the end of the contact breaker cam with the nylon heel of the points set, in each case.

5 It is sometimes found that there is a discrepancy between the points gaps of the 6CA unit when the scribe mark of the cam is aligned with the nylon heels. If the discrepancy is greater than 0.003 it is probably caused by cam run-out and can be cured by tapping the cam with a soft metal drift until it seats correctly. Cases have also occurred where the edge of one of the secondary backplates has fouled the cam. Contact between the cam and the backplate can result in the automatic advance unit remaining in the permanently retarded position, so if run-out is evident, either of these two faults should be investigated and remedied.

5 Contact breaker points - removal, renovation and replacement

1 If the contact breaker points are burned, pitted or worn, they must be removed for dressing. If, however, it is necessary to remove a substantial amount of material before the faces can be restored, the points should be renewed.

2 To remove the contact breaker points from the 4CA unit, remove the securing nuts from the condenser terminals. This will free the return springs from the moving contacts, which can be withdrawn from their respective pivot pins. Removal of the slotted nuts will free the fixed contact points.

3 In the case of the 6CA and 10CA unit, the moving points are removed by unscrewing the nut that secures each low tension lead wire and removing the lead and nylon bush. The return spring and contact point can then be withdrawn from each pivot pin. The fixed points are each retained by two screws which secure them to the backplate assembly.

4 The points should be dressed with an oilstone or fine emery cloth, taking care to keep them absolutely square throughout the dressing operation. If this precaution is not observed, the points will make angular contact with one another when they are replaced, and will burn away rapidly.

5 Replace the points by reversing the dismantling procedure. Take particular care to replace the insulating washers and spacers in their original positions, otherwise the points will be isolated electrically and the ignition system will no longer function.

6 It is important that the points are maintained in a clean condition, especially in machines fitted with provision for emergency start. The emergency start function depends on very small currents passing across the points and if a resistance builds up, the emergency start system will not function.

7 The timing side (right hand) cylinder ignition system is operated by the contact breaker to which the black/yellow lead is attached (left hand set of points, viewed from the open end of the contact breaker housing).

6 Condensers - removal and replacement

1 As mentioned in the preceding section, the condensers used in conjunction with the Lucas 4CV contact breakers assembly are contained within the contact breakers housing, attached to the backplate. It is advisable to withdraw the assembly complete before detaching the condensers, by removing the two pillar bolts securing the backplate to the timing cover housing.

2 When the Lucas 6CV contact breaker assembly is fitted, the condensers are located remotely. They are attached to a plate suspended from the bracket to which the front petrol tank mountings are bolted, immediately below the nose of the petrol tank.

3 If the engine becomes difficult to start, or if misfiring occurs, it is probable that a condenser has failed. Note that it is rare for both condensers to fail simultaneously, unless they have been damaged in an accident. Examine the contact breaker points whilst the engine is running to see whether arcing is taking place and, when the engine is stopped, examine the faces of the points. Arcing taking place or the points having a blackened and burnt appearance is characteristic of condenser failure.

4 It is not possible to check the condenser without the necessary test equipment. It is therefore best to fit a replacement condenser and observe the effect on engine performance, especially in view of the low cost of the replacement.

7 Ignition timing - checking and resetting

1 One of two methods can be used to check and if necessary reset the ignition timing, depending on whether a stroboscope is available or whether the static method has to be substituted in its absence. Although the stroboscope method is undoubtedly the more accurate, it is unlikely that many owners will have access to a stroboscope and the various attachments. In consequence, both methods are described in detail, so that the user of this workshop manual will have the option of following whichever the more appropriate.

2 If the static timing method is the one selected, commence by removing all four rocker box inspection caps (or finned inspection covers) and both spark plugs. Place the machine on the centre stand and engage top gear, so that the engine can be rotated by means of the rear wheel.

3 Arrange the right hand cylinder so that the piston is at top dead centre (TDC) on the compression stroke. It is best to use either a dial gauge pressing on the piston crown or a degree disc and adaptor shaft attached to the centre of the exhaust camshaft

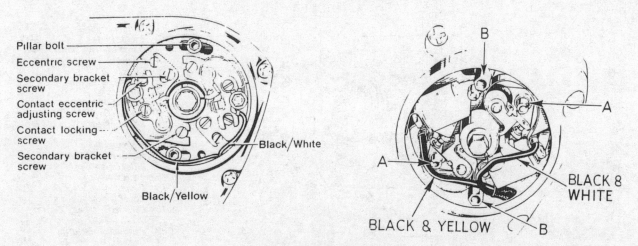

FIG.5.1. CONTACT BREAKERS

Contact breaker 6CA & 10CA   *Contact breaker 4CA*
To adjust contact breaker gaps slacken sleeve nuts 'A'. To rotate contact breaker base plate for setting ignition timing slacken pillar bolts 'B'.

(Triumph service tool D605/8 and nut S1-51) to maintain absolute accuracy of setting. In this latter case, a timing stick must be used to zero the timing disc accurately. Insert the stick through the right hand spark plug hole and make a mark on both the stick and an adjoining portion of the engine that coincide exactly when the piston is at TDC. Attach a pointer to some convenient part of the engine and rotate the timing disc, INDEPENDENT OF THE ENGINE, so that the zero mark aligns with the pointer. Make a second mark on the timing stick about 1 inch above the original, re-insert the stick and rotate the engine with timing disc attached until this second mark corresponds exactly with the mark originally made on the engine. Take a reading from the timing disc, at the pointer. Now rotate the engine in the opposite direction until the second mark on the timing stick again aligns with the mark on the engine, and take a further reading from the timing disc, at the pointer. If the two readings agree, the timing disc is zeroed correctly and the piston is exactly at TDC. Note that the timing stick should have a rubber plug or some similar device attached to prevent it from falling into the engine if the engine is inadvertently rotated too far in one direction.

4   This procedure can be short cut, on all engines numbered after DU 13375. These later engines have a plug on top of the crankcase, immediately to the rear of the cylinder barrel, which when removed, exposes the rim of the centre flywheel. If Triumph workshop tool D571/2 is screwed into the hole vacated by the plug, the inner plunger will drop into a hole drilled in the flywheel rim, when the piston is exactly at top dead centre. If the workshop tool is not available, the same effect can be achieved by placing a small socket spanner in the hole and using the shank of a drill as the plunger.

5   Check with the specifications section of this Chapter to ascertain the correct fully-advanced ignition setting recommended for the machine. Note the range of the automatic advance mechanism that is stamped on the REAR of the assembly, double this figure and subtract it from the full advance setting recommended for the machine. This is the correct STATIC setting for the engine. Convert this figure (expressed in degrees) to the equivalent piston displacement before TDC, if a timing stick is used WITHOUT a degree disc. If a degree disc is used, the static setting in degrees can be used without need for conversion.

6   Rotate the engine BACKWARDS to beyond the setting, then turn it forwards again until the correct setting is achieved. This procedure is necessary to take up any backlash that may otherwise affect the accuracy of the setting. Since the left hand

7.4 Using the crankcase timing plug to find top dead centre

set of contact breaker points (black and yellow lead) corresponds with the right hand cylinder, maintain the engine in the predetermined position and rotate the backplate of the contact breaker assembly (after slackening the pillar bolts) until the points are just on the point of separation. If the ignition is switched on during this operation, the exact point can be determined when the ammeter reading kicks back to zero. Tighten the pillar bolts.

7   Rotate the engine through 360 degrees, so that the left hand piston is now at TDC on the compression stroke. Check whether the right hand contact breaker points (black and white lead) are now in a similar position ie about to separate. If not, the accuracy of the second set of points must be corrected by adjusting the points gap. Minor adjustments are permissible; if more than 0.003 in adjustment is necessary, check for run-out as described in Section 4.5.

8   If a stroboscope is available, fit the timing shaft adaptor and timing disc to the exhaust camshaft and set the right hand piston to top dead centre as described in paragraphs 3 and 4 of this section (all engine numbers prior to DU 66245). Connect the stroboscope to the right hand plug lead, after the spark plugs and rocker inspection caps have been replaced, and start the

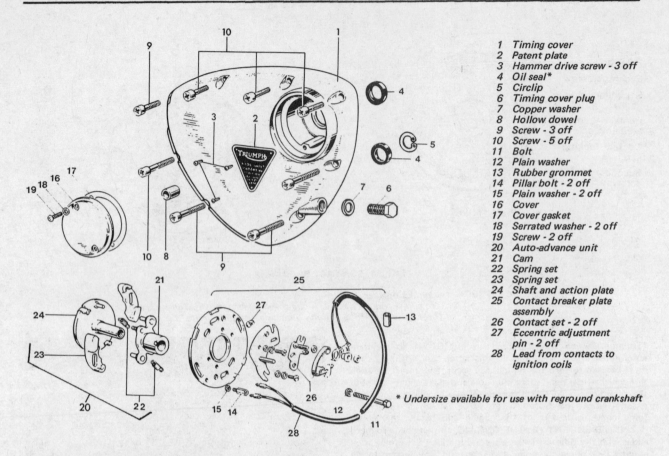

1	Timing cover
2	Patent plate
3	Hammer drive screw - 3 off
4	Oil seal*
5	Circlip
6	Timing cover plug
7	Copper washer
8	Hollow dowel
9	Screw - 3 off
10	Screw - 5 off
11	Bolt
12	Plain washer
13	Rubber grommet
14	Pillar bolt - 2 off
15	Plain washer - 2 off
16	Cover
17	Cover gasket
18	Serrated washer - 2 off
19	Screw - 2 off
20	Auto-advance unit
21	Cam
22	Spring set
23	Spring set
24	Shaft and action plate
25	Contact breaker plate assembly
26	Contact set - 2 off
27	Eccentric adjustment pin - 2 off
28	Lead from contacts to ignition coils

* Undersize available for use with reground crankshaft

FIG. 5.2. TIMING COVER AND CONTACT BREAKER

engine. Note the reading from the timing disc by shining the stroboscope light on the pointer, rewing the engine until full auto-advance is obtained. If the reading is not correct, adjust the contact breaker backplate on its slots (after slackening the pillar bolts) until the correct reading is achieved. Repeat this procedure for the left hand cylinder and make any adjustments by varying the contact breaker gap and NOT the backplate.

9 Before disconnecting the stroboscope and timing disc, a check can be made at lower engine speeds to verify whether the full range of ignition advance is achieved with both cylinders. It should be emphasised that the stroboscope method of engine timing ensures both plugs fire at exactly similar piston movements, thereby ensuring optimum engine running conditions and minimum vibration.

10 Machines having engines numbered above DU 66245 require a slightly different technique for stroboscope timing, although the setting up procedure is identical with that given in paragraph 4. These machines have an inspection plate fitted to the primary chaincase cover, retained by three screws. When the screws are withdrawn and the cover lifted away, it will be observed that there is a mark on the face of the generator rotor which should coincide with a pointer on the chaincase if the ignition timing setting is correct. If the machine does not have a chaincase pointer, it is necessary to fit temporarily a special timing plate (Triumph part number D2014) in its place. The plate has two markings; only the line marked 'B' is applicable to the larger capacity models.

11 Connect the stroboscope to the right hand spark plug lead and start the engine. Shine the stroboscope lamp on the rotor marking in the vicinity of the chaincase pointer or plate with the engine running at more than 2000 rpm. Adjust the contact breaker backplate until the marks coincide, when the timing of

the right hand cylinder is correct. Connect the stroboscope to the left hand cylinder and repeat the procedure, in this case turning the eccentric screw in the right hand set of contact breaker points until the marks coincide.

12 When the timing has been verified as correct, remove the various attachments and replace the end cover of the contact breaker assembly, using a new gasket. If the stroboscope method of timing has been used, remove the timing plate from the primary chaincase and replace the circular cover. Replace the rocker box inspection caps or the finned inspection covers.

13 Note that if a 6 or 12 volt stroboscope is used, the power must be from an external source. If the machine's battery is used, there is risk of AC pulses developing in the low tension circuit that may cause the stroboscope to give a false reading.

8 Automatic advance unit - removal, examination and replacement

1 Fixed ignition timing is of little advantage as the engine speed increases and it is therefore necessary to incorporate a method of advancing the timing by centrifugal means. A balance weight assembly located behind the contact breaker, linked to the contact breaker cam, is employed in the case of the Triumph unit-construction twins. It is secured to the exhaust camshaft by a bolt that passes through the centre of the contact breaker cam. It can be withdrawn without need to remove the timing cover, if the contact breaker assembly is removed first.

2 When the assembly is removed from the machine, it is advisable to make a note of the degree figure stamped on the back of the cam unit. This relates to the ignition advance range and, as the previous section has indicated, must be known for accurate static timing.

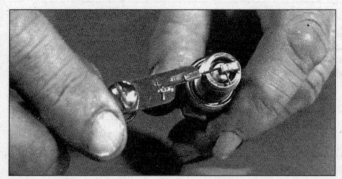

Electrode gap check - use a wire type gauge for best results

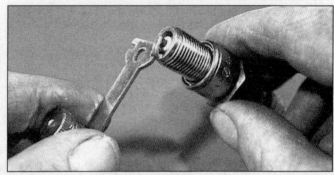

Electrode gap adjustment - bend the side electrode using the correct tool

Normal condition - A brown, tan or grey firing end indicates that the engine is in good condition and that the plug type is correct

Ash deposits - Light brown depcsits encrusted on the electrodes and insulator, leading to misfire and hesitation. Caused by excessive amounts of oil in the combustion chamber or poor quality fuel/oil

Carbon fouling - Dry, black sooty deposits leading to misfire and weak spark. Caused by an over-rich fuel/air mixture, faulty choke operation or blocked air filter

Oil fouling - Wet oily deposits leading to misfire and weak spark. Caused by oil leakage past piston rings or valve guides (4-stroke engine), or excess lubricant (2-stroke engine)

Overheating - A blistered white insulator and glazed electrodes. Caused by ignition system fault, incorrect fuel, or cooling system fault

Worn plug - Worn electrodes will cause poor starting in damp or cold weather and will also waste fuel

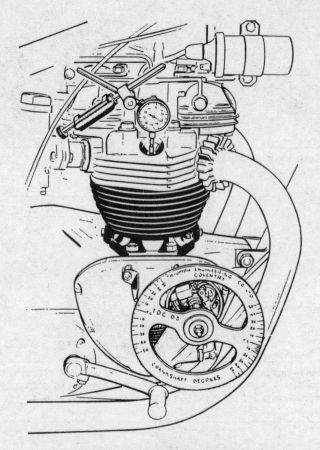

Fig. 5.3. Engine fitted with timing disc and dial indicator

3 The unit is most likely to malfunction as the result of condensation, which will cause rusting to take place. This will immediately be evident when the assembly is removed. Check that the balance weights move quite freely and that the return springs are in good order. Before replacing the assembly by reversing the dismantling procedure, lubricate the balance weight pivot pins and the cam spindle, and place a light smear of grease on the face of the contact breaker cam. Lubricate the felt pad that bears on the contact breaker cam.

9 Ignition cut-out

1 The ignition circuit is controlled by the ignition switch on the left hand top fork cover. The key can only be withdrawn from the switch when it is in the 'OFF' position and the ignition circuit broken.
2 A state of emergency can occur when the machine is on the move and it is not convenient to reach out for the switch key. Late models have an additional cut-out button on the left hand side of the handlebars, which will break the ignition circuit all the time it is depressed. On these models, the separate lighting switch is mounted in the headlamp shell. Because the ignition circuit is broken only when the cut-out button is depressed, it is essential to turn the switch to the off position and remove the key when the machine is parked.

10 Spark plugs - checking and resetting the gap

1 A 14 mm spark plug is fitted to both cylinders of the Triumph unit-construction twins, the grade depending on the model designation. Refer to the Specifications of this Chapter for the recommended grades.
2 All 650/750 cc models use spark plugs with a ¾ in reach which should be gapped to the figure given in the Specifications at the beginning of this Chapter. Always use the grade of plug recommended or the exact equivalent in another manufacturer's range.
3 Check the gap at the plug points every 3000 miles or during every six monthly service, whichever is soonest. To reset the gap bend the outer electrode to bring it closer to the inner electrode and check that the appropriate size feeler gauge can just be inserted. Never bend the centre electrode otherwise the insulator will crack, causing engine damage if particles fall in whilst the engine is running.
4 With some experience, the condition of the spark plug electrodes and insulators can be used as a reliable guide to engine operating conditions. See the accompanying illustrations for examples.
5 Always carry a pair of spark plugs of the correct grade. In the rare event of plug failures, they will enable the engine to be restarted.
6 Never overtighten a spark plug, otherwise there is risk of stripping the threads from the cylinder head, particularly those cast in light alloy. The plugs should be sufficiently tight to seat firmly on their sealing washers. Use a spanner that is a good fit, otherwise the spanner may slip and break the insulator.
7 Make sure the plug caps are a good fit and free from cracks. These caps contain the suppressor that eliminates radio and TV interference.

11 Fault diagnosis

Symptom	Reason/s	Remedy
Engine will not start	No spark at plug	Check whether contact breaker points are opening and also whether they are clean. Check wiring for break or short circuit.
Engine fires on one cylinder	No spark at plug or defective cylinder	Check as above, then test ignition coil. If no spark, see whether points arc when separated. If so, renew condenser.
Engine starts but lacks power	Automatic advance unit stuck or damaged	Check unit for freedom of action and broken springs.
	Ignition timing retarded	Verify accuracy of timing. Check whether points gaps have closed.
Engine starts but runs erratically	Ignition timing too far advanced	Verify accuracy of timing. Points gaps too great.
	Spark plugs too hard	Fit lower grade of plugs and retest.

Chapter 6 Frame and forks

Contents

Specifications

Oil tank capacity:

Oil tank models	5 Imp pints (2.84 litre)
Oil in frame models	4 Imp pints (2.27 litre)

Front forks

	Free length – new	Colour code
Springs – 1963 models (DU101 to DU5824):		
6T, T120 solo	17.06 in (433.32 mm)	Black/Green
6T, T120 sidecar – with longer fork lower legs	18.31 in (465.07 mm)	Red/White
TR6 solo	N/Av	Black/White
TR6 sidecar	N/Av·	Black/Red
Springs – 1964 models (DU5825 to DU13374):		
Solo	8.75 in (222.25 mm)	Unpainted
Sidecar	8.75 in (222.25 mm)	Yellow/White
Springs – 1965 to 1970 models:		
Solo – except T120TT	9.75 in (247.65 mm)	Yellow/Blue
Sidecar and T120TT	9.75 in (247.65 mm)	Yellow/Green
Springs – 1971 on (solo only)	19.10 in (485.14 mm)	Orange
Oil capacity – per leg:		
1963 models solo	$\frac{1}{4}$ pint (150 cc) SAE 20W/30	
1963 models sidecar – with longer fork lower legs ...	$\frac{3}{8}$ pint (225 cc) SAE 20W/30	
All other models...	$\frac{1}{3}$ pint (190 cc) SAE 20	

1 General description

A full cradle frame is fitted to the Triumph unit-construction twins, in which the front down tube branches into two duplex tubes at the lower end which form the cradle for the unit-construction engine/gear unit.

Rear suspension is provided by a swinging arm assembly that pivots from a lug welded to the vertical tube immediately to the rear of the gearbox. Movement is controlled by two hydraulically-damped rear suspension units, one on each side of the sub-frame. The units have three-rate adjustment, so that the spring loading can be varied to match the conditions under which the machine is to be used.

Front suspension is provided by telescopic forks of conventional design.

Extensive redesigning took place during 1970 and from the 1971 models onward a different frame of the duplex tube type was substituted. The most distinctive feature was the use of an enlarged vertical seat pillar to contain the engine oil in place of the hitherto side-mounted oil tank. Front suspension was also changed. Forks of the 'slimline' type replaced the originals, having slim, exposed stanchions with internal springs and a different damper assembly.

2 Front forks - removal from frame

1 It is unlikely that the front forks will need to be removed from the frame as a complete unit unless the steering head bearings require attention or the forks are damaged in an accident.

2 Commence operations by placing the machine on the centre stand and disconnecting the front brake. If the split pin and clevis pin through the U shaped connection to the brake operating arm are removed, the cable can be pulled clear of the cable stop and removed from the cable guide attached to the front mudguard.

3 To remove the front wheel, unscrew and remove the two bolts securing the lower half of each split clamp to the bottom of the fork legs (four nuts, late models). When these clamps are withdrawn, the wheel will drop clear, complete with brake plate and wheel spindle. It may be necessary to raise the front end of the machine a little, in order to gain sufficient clearance for the wheel to clear the mudguard when it is removed.

4 It is convenient at this stage to drain the fork legs of their oil content, if the fork legs are to be dismantled at a later stage. The drain plug is found above the wheel spindle recess on each fork leg. Remove both drain plugs and leave the forks to drain into some suitable receptacle whilst the dismantling continues.

5 There is no necessity to remove the front mudguard unless the fork legs are to be dismantled. The lower mudguard stays bolt direct to lugs at the lower end of each fork leg; the centre fixing is made to the inside of each fork leg where a shaped lug accepts the cut-out of the mudguard stay assembly, which is then retained by a bolt and washer. If the various nuts, bolts and washers are removed, the mudguard can be withdrawn, complete with stays.

6 Detach the headlamp after disconnecting the battery leads.

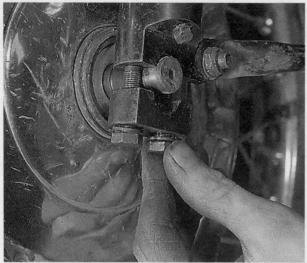

2.3 Front wheel spindle is retained by split clamps

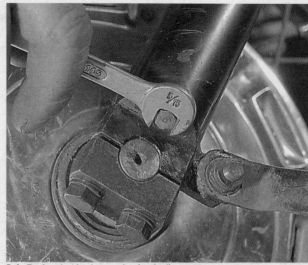

2.4 Drain plug is above wheel spindle recess

2.11a Remove plated fork top nuts to release inner stanchions

2.11b Rubber gaiters are retained by wire clips

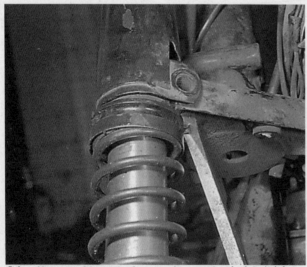

3.1a Use screwdriver to spring open lower yoke pinch bolt joint

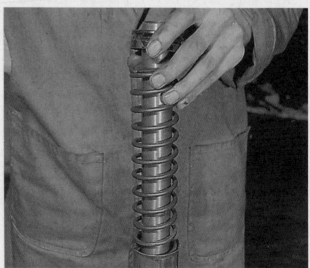

3.1b Fork spring is exposed after gaiter is removed

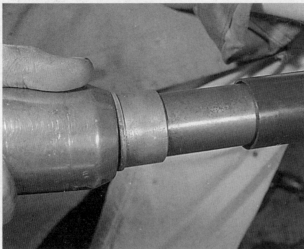

3.2 Removal of dust excluder sleeve frees stanchion from lower fork leg

3.3 Stanchion will free from lower fork leg with a sharp pull

3.5 Circlip prevents shuttle valve from passing into stanchion

Commence by slackening the screw at the top of the headlamp rim, which will allow the rim and reflector unit to be removed. Detach the pilot bulb holder and pull the snap connectors from the main bulb holder. Disconnect the four spade terminals from the lighting switch and from the ammeter. On earlier models, snap connectors at the wiring harness form an alternative means of disconnection. Disconnect the snap connectors in the warning light leads, then withdraw the wiring harness complete from the headlamp shell. Note that the harness will bring with it the warning light bulb holders and rubber grommets.

7 Remove the pivot bolts on each side of the headlamp shell and withdraw the shell, complete with spacers. A slightly different procedure is necessary in the case of models fitted with flashing indicators, the arms of which pass through the headlamp shell to form the pivots. In this case it is necessary to disconnect the indicator leads and unscrew the nuts around the extensions of the indicator arms which project through the headlamp shell to hold it in position.

8 Detach the control cables from the handlebar controls, or the controls themselves, complete with cables. Remove the dip-switch, the cut-out button (if fitted) and the horn push. Remove the ignition switch, retained in the left hand top fork cover extension by a locknut. The handlebars can now be removed by unscrewing the two eye bolts underneath the top yoke of the forks, or by removing the split clamps if the handlebars are not rubber-mounted.

9 Remove the steering damper knob and rod (if fitted) by unscrewing the knob until the rod is released from the lower end into which it threads. Slacken the pinch bolt through the top fork yoke, found at the rear of the steering assembly, above the tank. Unscrew the fork stem sleeve nut (if steering damper fitted) or the blind nut (domed) at the top of the steering head column. Slacken and remove the two plated nuts at the top of each fork leg that pass through the top fork yoke.

10 Remove the top fork yoke by tapping on the underside with a rawhide mallet. The forks should be supported throughout this operation because they will free immediately the yoke clears the tapers of the fork inner tubes or stanchions. When the yoke is displaced and removed, the complete fork assembly can be lowered from the steering head and drawn clear. Note that as the head races separate the uncaged ball bearings will be released. Arrangements should be made to catch the ball bearings as they drop free; most probably only those from the lower race will be displaced.

11 It is possible to remove the fork legs separately, if there is no reason to disturb either the steering head assembly or the fork yokes. In this case the plated top fork nuts should be removed and the pinch bolts through each side of the lower fork yoke slackened and removed. Triumph service tool Z169 (Z19 for machines with engine numbers prior to DU 68363) should then be threaded into the top of each stanchion to the full depth of thread and used as a drift to drive each stanchion taper free, to release the stanchion complete with lower leg, so that it can be passed through the lower fork yoke and withdrawn from the machine. It is often necessary to open up the pinch bolt joint in the lower fork yoke to prevent the stanchion from binding. If the stanchion has rusted, this will impede its progress through the bottom yoke. Remove all surface rust with emery cloth, wipe clean and apply a light coating of grease or oil.

12 It must be emphasised that the use of the recommended Triumph service tool is essential for this operation. The stanchions are a very tight fit and any attempt to free them with a punch or drift will invariably damage or distort the internal threads, necessitating replacement. An old top fork nut can sometimes be used successfully as a substitute but only if the stanchion is not a particularly tight fit, in either of the fork yokes.

3 Front forks - dismantling, examining and reassembling the fork legs

1 When the legs have been removed from the machine, withdraw the fork top covers, or, in the case of some of the early models, the nacelle bottom covers. Remove the cork seating washers. If rubber gaiters are fitted, the top and bottom retaining clips

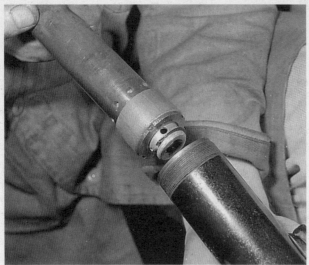

3.9 Lower bush is retained by sleeve nut

6.5 Do not forget to add damping oil before replacing fork top nuts

8.1a Slacken bolt behind left rear suspension unit to free end of chainguard

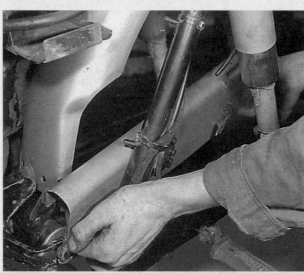

8.1b Front of chainguard is attached to swinging arm cross-member

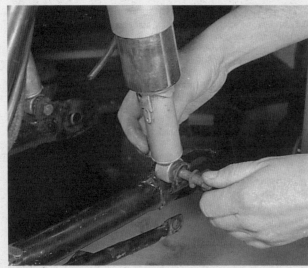

8.2 Disconnect rear suspension units at lower ends only

8.3a Tab washer locks swinging arm pivot nut

should be slackened and the gaiters removed, together with the clips. Remove the spring abutments and the fork springs.

2 Removal of the dust excluder sleeves which contain a plain washer and oil seal is facilitated by the use of Triumph service tool D220, which engages with the peg holes in the outer surface. If the service tool is not available, a strap spanner or careful work with a centre punch will provide an alternative solution. The fork leg should be supported in a vice during this operation, by clamping the wheel spindle recess. A sharp knock is needed to free the sleeve initially; thereafter it can be unscrewed and removed.

3 Withdraw the stanchion from each leg by withdrawing it whilst the fork leg is still clamped in the vice. A few sharp pulls may be necessary to release the top bush which is a tight fit in the fork leg.

3 If the frame number of the machine is within the range DU 5825 - DU 66245, a different type of internal hydraulic damper assembly was fitted. Before the stanchions can be freed from the fork legs, it is necessary to first unscrew the hexagon headed bolt counter bored into the wheel spindle recesses.

4 Damper units fitted to forks used in conjunction with frame numbers prior to DU 5824 employ an oil restrictor rod assembly, secured by a bolt counter bored into the front wheel spindle recesses of the lower fork legs. Unlike the design that followed, it is not necessary to detach this assembly before the stanchions can be withdrawn from the lower legs.

5 A further change of damper unit occurred with forks fitted to frames numbered DU 66245 onward. These forks are fitted with a shuttle valve damper, attached to the lower end of the fork stanchions. Each shuttle valve is retained by a sleeve nut which also holds the bottom bearing of the fork leg. A circlip in front of this nut prevents the valve from passing into the stanchion. This type of damper assembly is easily recognisable by the eight oil bleed holes above the bottom bearing location.

6 The parts most liable to become damaged in an accident are the fork stanchions, which will bend on heavy impact. To check for misalignment, roll the stanchion on a sheet of plate glass, when any irregularity will be obvious immediately. It is possible to straighten a stanchion that has bowed not more than 5/32 in out of true but it is debatable whether this action is desirable. Accident damage often overstresses a component and because it is not possible to determine whether the part being examined has suffered in this way, it would seem prudent to renew rather than repair.

7 Check the top and bottom fork yokes which may also twist or distort in the event of an accident. The top yoke can be checked by temporarily replacing the stanchions and checking whether they lay parallel to one another. Check the lower fork yoke in the same manner, this time with the stanchions inserted until about 6½ inches protrude. Tighten the pinch bolts before checking whether the stanchions are parallel with one another. The lower yoke is made of a malleable material and can be straightened without difficulty or undue risk of fracture.

8 It is possible for the lower fork legs to twist and this can be checked by inserting a dummy wheel spindle made from 11/16 inch diameter bar and replacing the split retaining clamps. If a set square is used to check whether the fork leg is perpendicular to the wheel spindle, any error is readily detected. Renewal of the lower fork leg is necessary if the check shows misalignment.

9 The fork bushes can be checked by positioning the top bush close to the lower bush at the bottom of the stanchion and inserting the assembly in the lower fork leg. Any undue play will immediately be evident, necessitating renewal of the bushes. If the forks are fitted with the older type of grey sintered iron bushes they should be replaced by the later sintered bronze bushes, to gain the benefit of a reduced rate of wear.

10 Examine both fork springs and check that they are of the same length and have not compressed. If either has settled to a length of ½ in/12.7 mm (models with long internal springs) or ¼ in/6.35 mm (models with short external springs) or more, shorter than that specified, both springs must be renewed as a pair. Both springs must be of the same colour-coding and of the

correct type for the machine and its intended use.

11 Although the post 1970 'slimline' forks are different in appearance, the same broad dismantling procedure applies. If difficulty is encountered in removing the oil seals, they can be levered out of position by making up a special tool, as shown in the accompanying drawing. This will obviate risk of damage to the oil seal housing. The tool must not contact the oil seal housing and should be worked around the periphery of the seal otherwise it will rip through the sealing lip rather than displace the seal.

4 Front forks - examining the steering head races

1 If the steering head races have been dismantled, it is advisable to examine them prior to reassembling the forks. Wear is usually evident in the form of indentations in the hardened cups and cones, around the ball track. Check that the cups are a tight fit in the steering column headlug.

2 If it is necessary to renew the cups and cones, use a drift to displace the cups by locating with their inner edge. Before inserting the replacements, clean the bore of the headlug. The replacement cups should be drifted into position with a soft metal drift or even a wooden block. To prevent misalignment, make sure that the cups enter the headlug bore squarely. The lower cone can be levered off the bottom fork yoke with tyre levers; the upper cone is within the top fork yoke and can be drifted out. Clean up any burrs before the new replacements are fitted. A length of tubing which will fit over the head stem can be used to drive the lower cone into position so that it seats squarely.

3 When the cups and cones are replaced, discard the original ball bearings and fit a new set. It is false economy to re-use the originals in view of the very low renewal cost. Forty ¼ inch diameter balls are required, 20 for each race. Note that when the bearing is assembled, the race is not completely full. There should always be space for one bearing, to prevent the bearings from skidding on one another and wearing more rapidly. Use thick grease to retain the ball bearings in place whilst the forks are being offered up.

5 Front forks - reassembling the fork legs

1 The fork legs are reassembled by following the dismantling procedure in reverse. Make sure all of the moving parts are lubricated before they are assembled. Fit new oil seals, regardless of the condition of the originals. When refitting the oil seals or new replacements, place a small polythene bag over the end of a stanchion and slide the oil seal downwards. Place the stanchion in the lower fork leg and use it to position the oil seal correctly, whilst the latter is pressed into place. The polythene bag prevents the sharp lip at the edge of the stanchion from damaging the precision edge of the seal.

2 Before refitting the fork stanchions, make sure the external surfaces are clean and free from rust. This will make fitting into the fork yokes at a later stage much easier. Oil, or lightly grease, the outer surfaces after removing all traces of the emery cloth or other cleaner used.

3 When refitting the fork gaiters (if originally fitted) check that they are positioned correctly. The small hole near the area where the clip fastener is located should be at the bottom.

6 Front forks - refitting to frame

1 If it has been necessary to remove the fork assembly complete from the frame, refitting is accomplished by following the dismantling procedure in reverse. Check that none of the ball bearings are displaced whilst the steering head stem is passed through the headlug; it has been known for a displaced ball to fall into the headlug and wear a deep groove around the headstem of the lower fork yoke.

2 Take particular care when adjusting the steering head bearings. The blind or sleeve nut should be tightened sufficiently to remove all play from the steering head bearings and no more. Check for play by pulling and pushing on the fork ends and make sure the handlebars swing easily when given a light tap on one end.

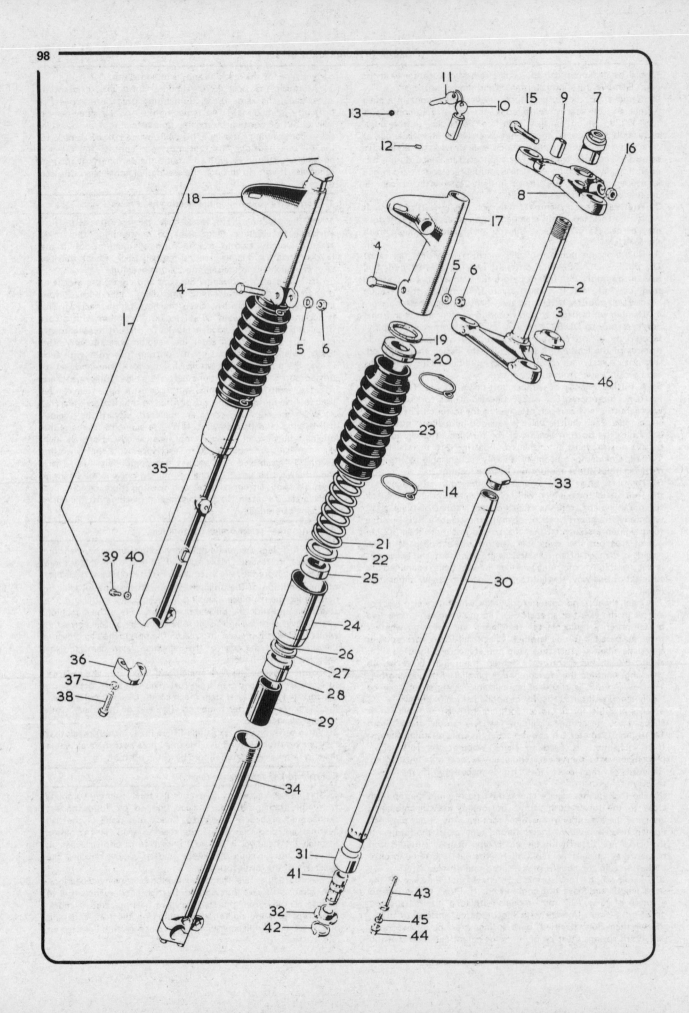

FIG. 6.1. TELESCOPIC FRONT FORKS (PRE-1970)

1	Fork assembly complete		24	Dust excluder sleeve - 2 off
2	Bottom fork yoke and headstem		25	Oil seal - 2 off
3	Bottom steering head cone		26	'O' ring - 2 off
4	Pinch bolt - 2 off		27	Plain washer - 2 off
5	Plain washer - 2 off		28	Top bearing - 2 off
6	Nut - 2 off		29	Damping sleeve - 2 off
7	Sleeve nut		30	Stanchion - 2 off
8	Top fork yoke		31	Lower bearing - 2 off
9	Bonded bush - 2 off		32	Bearing nut - 2 off
10	Lock complete with two keys		33	Top fork nut - 2 off
11	Key		34	Lower fork leg complete with cap (left)
12	Grub screw		35	Lower fork leg complete with cap (right)
13	Sealing washer		36	Spindle cap - 2 off
14	Gaiter clip - 4 off		37	Spring washer - 4 off
15	Pinch bolt		38	Cap bolt - 4 off
16	Seated nut		39	Drain plug - 2 off
17	Top fork cover (left)		40	Fibre washer - 2 off
18	Top fork cover (right)		41	Shuttle valve - 2 off
19	Cork washer - 2 off		42	Circlip - 2 off
20	Spring abutment		43	Restrictor - 2 off
21	Fork spring - 2 off		44	Aluminium washer
22	Plain washer - 2 off		45	Flanged bolt - 2 off
23	Fork gaiter - 2 off		46	Locating pin

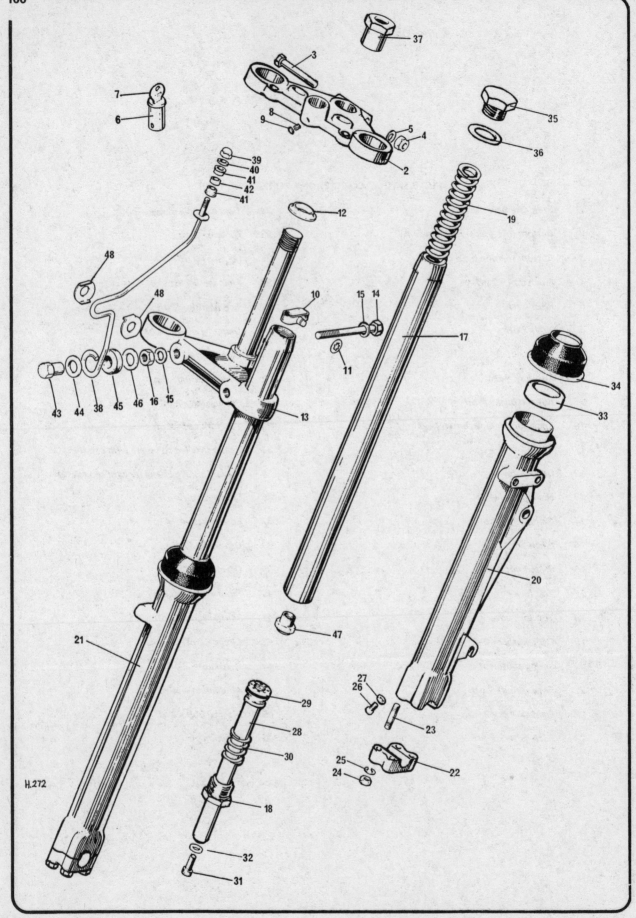

H.272

FIG. 6.2. TELESCOPIC FRONT FORKS (POST-1970)

1	Fork assembly complete	25	Washer - 8 off
2	Fork top yoke	26	Drain plug - 2 off
3	Pinch bolt	27	Washer - 2 off
4	Nut	28	Damper tube and valve assembly - 2 off
5	Washer	29	'O' ring, damper valve - 2 off
6	Lock, complete with 2 keys	30	Recoil spring - 2 off
7	Key	31	Cap screw, damper tube - 2 off
8	Grub screw	32	Cap screw seal - 2 off
9	Sealing washer	33	Oil seal outer member - 2 off
10	Brake cable retainer	34	Scraper sleeve outer member - 2 off
11	Starlock washer	35	Top fork nut - 2 off
12	Bottom steering head cone	36	Washer - 2 off
13	Bottom fork yoke steering head stem	37	Steering head nut
14	Pinch bolt - 2 off	38	Headlamp bracket left and right
15	Washer - 4 off	39	Nut - 2 off
16	Nut - 2 off	40	Washer - 2 off
17	Stanchion - 2 off	41	Bush - 2 off
18	End plug - 2 off	42	Spacer - 2 off
19	Fork spring - 2 off	43	Sleeve nut - 2 off
20	Lower fork leg (left)	44	Washer - 2 off
21	Lower fork leg (right)	45	Grommet - 2 off
22	Wheel spindle cap - 2 off	46	Washer - 2 off
23	Stud - 8 off	47	Nut, damper valve - 2 off
24	Nut - 8 off	48	Clamp washer - 4 off

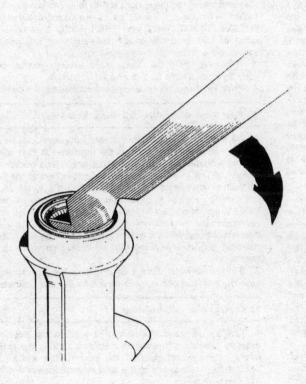

FIG. 6.3 . REMOVING OIL SEAL

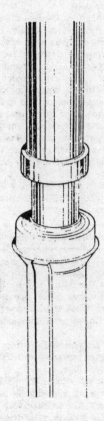

FIG. 6.4. REPLACING OIL SEAL

3 It is possible to overtighten the steering head bearings and place a loading of several tons on them, whilst the handlebars appear to turn without difficulty. On the road, overtight head bearings cause the steering to develop a slow roll at low speeds.

4 Before the plated top fork nuts are replaced, do not omit to replace the drain plug in each fork leg and to refill each leg with the correct quantity of SAE 20 Oil. Each leg holds 190 cc

5 Difficulty will be experienced in raising the fork stanchions so that their end taper engages with the taper inside the top fork yoke. Triumph service tool Z170 (unified threads) or Z161 (CEI threads) is specified for this purpose; if the service tool is not available, a wooden broom handle screwed into the inner threads of the fork stanchion can often be used to good effect.

6 Before final tightening, bounce the forks several times so that the various components will bed down into their normal working locations. This same procedure can be used if the forks are twisted, but not damaged, as the result of an accident. Always retighten working from the bottom upward.

7 Frame assembly - examination and renovation

1 If the machine is stripped for a complete overhaul, this affords a good opportunity to inspect the frame for cracks or other damage which may have occurred in service. Check the front down tube immediately below the steering head and the top tube immediately behind the steering head, the two points where fractures are most likely to occur. The straightness of the tubes concerned will show whether the machine has been involved in a previous accident.

2 If the frame is broken or bent, professional attention is required. Repairs of this nature should be entrusted to a competent repair specialist, who will have available all the necessary jigs and mandrels to preserve correct alignment. Repair work of this nature can prove expensive and it is always worthwhile checking whether a good replacement frame of identical type can be obtained from a breaker or through the manufacturer's Service Exchange Scheme. The latter course of action is preferable because there is no safe means of assessing on the spot whether a secondhand frame is accident damaged too.

3 The part most likely to wear during service is the pivot and bush assembly of the swinging arm rear fork. Wear can be detected by pulling and pushing the fork sideways, when any play will immediately be evident because it is greatly magnified at the fork end. A worn pivot bearing will give the machine imprecise handling qualities which will be most noticeable when traversing uneven surfaces.

8 Swinging arm rear suspension - examination and renovation

1 If wear is evident in the swinging arm pivot, it will be necessary to remove the swinging arm fork. Commence by placing the machine on the centre stand and remove both the final drive chain and the rear wheel complete with sprocket, following the procedure detailed in Chapter 7.9. Remove the rear chainguard, the left hand cover containing the tool kit and both rear engine plates.

2 Disconnect the rear suspension units from the swinging fork ends by removing the nuts, bolts and washers securing the lower end of each unit to the fork lugs.

3 Remove the nut and tab washer from the pivot pin which passes through the frame lug immediately behind the gearbox and withdraw the pin itself. The swinging fork can now be pulled away from the machine, rearwards, after disconnecting the short rubber tube that forms part of the system for lubricating the final drive chain. The fork will withdraw complete with the bearing assembly and the flanged washers on each side of the pivot.

4 The two sleeve bushes, together with the distance tube between them, can be pushed out of the fork pivot, leaving only the pivot bushes in place, one at each end. Use a soft metal drift to displace the bushes, which are a good press fit in the pivot tube.

5 New bushes are of the steel backed type and if pressed in carefully, using a smear of grease to aid assembly, will give the correct working clearance without need for reaming. Check that the bushes have no burrs before reassembling with the sleeve bushes and distance tube.

6 Before replacing the swinging fork in the frame, check that it is straight and does not have a lateral twist. Pack the assembly with new grease, refit the flanged washers with new O ring internal seals and re-align the fork with the frame lug so that the pivot pin can be pushed home and the tab washer and nut refitted. Tighten the nut and bend the tab washer to lock it in place, then reconnect the ends of the rear suspension units. The remainder of the reassembly work should be accomplished by following the dismantling procedure, in reverse.

9 Rear suspension units - examination

1 Only a limited amount of dismantling can be undertaken because the damper unit is an integral part of each unit and is sealed. If the unit leaks oil, or if the damping action is lost, the unit must be replaced as a whole after removing the compression spring and outer shroud.

2 If the units require attention, place the machine on the centre stand and remove both nuts and bolts, so that each unit can be detached from the machine. The spring and outer shroud are removed by clamping the lower end of the unit in a vice and depressing the outer shroud so that the split collets which seat in the top of the shroud can be displaced. The shroud and spring can then be lifted off. Note that the suspension units should be set to their light load position, to make this task easier. The standard rating for the springs is 145 lb in and the fitted length 8 inches. The springs should be colour-coded blue/yellow. The R120TT, T120R, TR6C and TR6R models have lighter springs rated at 100 lb in having a fitted length of 8.4 inches. These springs are colour-coded green/green.

3 When replacing the units, make sure the rubber bushes in each 'eye' are a tight fit and in good condition.

4 Post 1970 models have exposed springs and no outer shroud.

10 Rear suspension units - adjusting the loading

1 As mentioned earlier in the text, the units can be adjusted without detaching them from the machine, so that the loading can be adjusted to that most suitable for the conditions under which the machine is to be used. A built-in cam at the lower end of each unit permits the sleeve carrying the lower end of the compression springs to be rotated, so that the sleeve is adjustable to different heights. The lowest position, or lightest loading is recommended for solo riding, the middle or medium loading for heaviest loading when a pillion passenger is carried. A special 'C' spanner, provided with the tool kit is used to effect the adjustments.

2 Both units must always be set to an identical rating, otherwise the handling of the machine will be seriously affected.

11 Centre stand - examination

1 The centre stand is attached to lugs on the bottom frame tubes to provide a convenient means of parking the machine or raising either wheel clear of the ground. The stand pivots on bolts through these lugs and is held retracted when not in use by a return spring.

2 The condition of the return spring and the return action of the stand should be checked regularly, also the security of the two nuts and bolts retained by a tab washer. If the stand drops whilst the machine is in motion, it may easily catch in some road obstacle and unseat the rider.

3 Some riders remove the centre stand completely because it is inclined to ground if the machine is cornered vigorously. It is questionable whether such action can be justified because in the event of a puncture, there is no alternative means of supporting the machine whilst either wheel is removed.

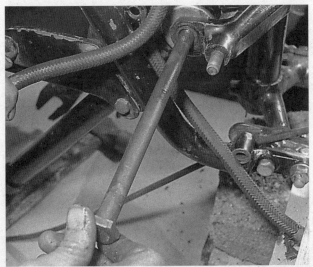

8.3b Withdraw pivot pin to release fork

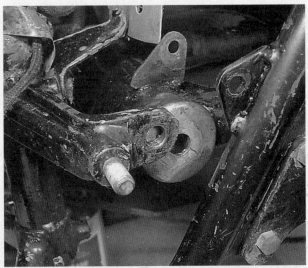

8.3c Fork will pull away complete with flanged washers

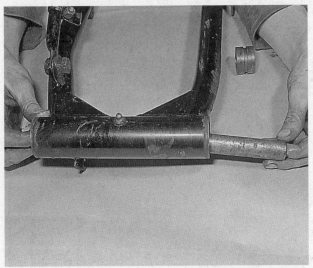

8.4 Sleeve bushes and distance piece will push out with ease

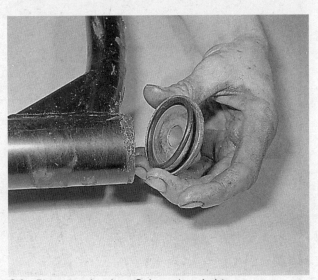

8.6a Flanged washers have O ring seals on inside

8.6b Pack bearing assembly with new grease

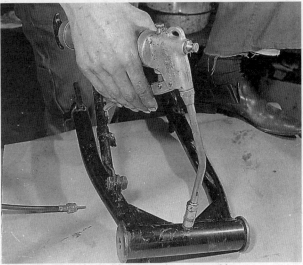

8.6c Fill whole pivot with grease prior to assembly

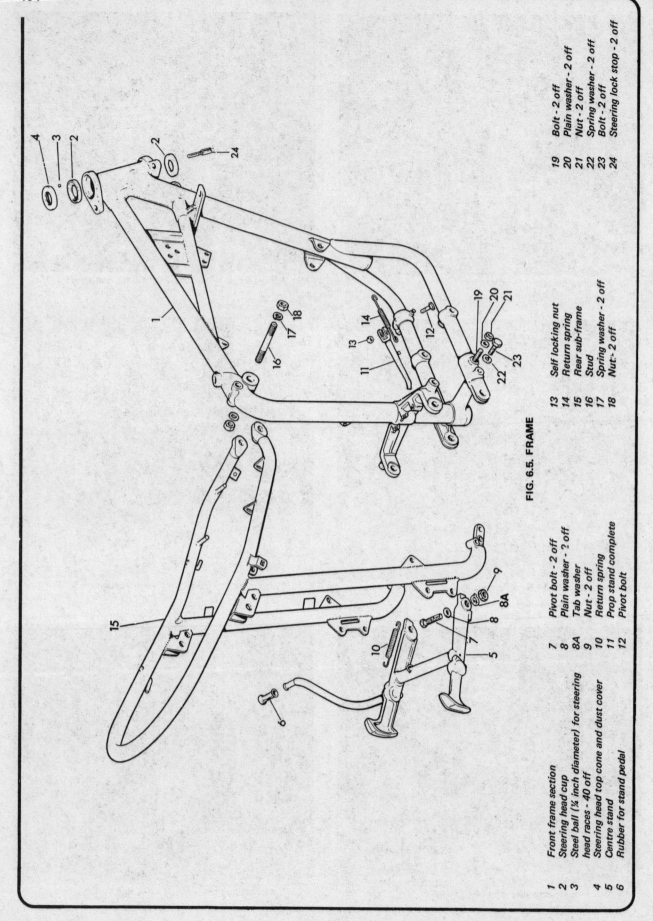

FIG. 6.5. FRAME

1	Front frame section
2	Steering head cup
3	Steel ball (¼ inch diameter) for steering head races - 40 off
4	Steering head top cone and dust cover
5	Centre stand
6	Rubber for stand pedal
7	Pivot bolt - 2 off
8	Plain washer - 2 off
8A	Tab washer
9	Nut - 2 off
10	Return spring
11	Prop stand complete
12	Pivot bolt
13	Self locking nut
14	Return spring
15	Rear sub-frame
16	Stud
17	Spring washer - 2 off
18	Nut - 2 off
19	Bolt - 2 off
20	Plain washer - 2 off
21	Nut - 2 off
22	Spring washer - 2 off
23	Bolt - 2 off
24	Steering lock stop - 2 off

12 Prop stand - examination

1 A prop stand which pivots from the left hand lower frame tube is provided for occasional parking, when it is not considered necessary to use the centre stand.

2 At regular intervals check that the prop stand return spring is in good order and that the pivot nut and bolt are not working loose.

13 Footrests and rear brake pedal - examination

1 The footrests are attached to the rear engine plates by a nut and washer behind each plate. They are prevented from rotating by two small pegs engaging with holes in the engine plates. Some of the earlier models have footrests which engage with mounting lugs below the engine unit.

2 If the machine is dropped, it is probable that the footrests will bend; they are quite soft. To straighten, they must be removed from the machine and their rubbers detached. They should be held in a vice during the straightening operation, using a blow lamp to heat the area where the bend occurs to a cherry red. If they are bent cold, there is risk of fracture.

3 The rear brake pedal passes through the rear engine plates where it engages with the lever carrying the brake rod. If the lever should bend, it can be straightened in a similar manner, after removal from the machine.

14 Oil tank - removal and replacement

1 Under normal circumstances, it is unlikely that the oil tank will need to be removed from the machine. It is rubber mounted, to isolate the tank from high frequency vibration which may cause the welded seams to split.

2 Drain the oil tank by removing the drain plug at the base of the tank or by pulling off the flexible main feed pipe, after slackening the retaining clip or unscrewing the union nut. Remove the filler cap, the flexible return feed pipe (after slackening the retaining clip), the flexible connection to the froth tower or filler orifice neck and the flexible connection at the T junction of the short return feed pipe used to form a convenient take-off for the rocker feed.

3 If a tool tray is fitted, this should be removed. It is retained by two nuts and bolts.

4 Disconnect the battery and remove the battery carrier, as described in Chapter 8, section 3. Remove both nuts from the oil tank mounting pegs, using a screwdriver to prevent the rubbers from turning. Removal of the forward mounted nut will free the red leads which earth the electrical system to the frame; removal of the rearmost nut releases the check strap which limits the swing of the dual seat when it is raised.

5 Remove the bolt retaining the oil tank bottom bracket and remove the bracket completely, taking care not to lose the mounting grommet.

6 Push the slotted pegs through the mounting rubbers, lift the tank and push the top inwards so that the froth tower clears the frame. Pull the lower end of the tank outward and when it is clear of the machine, withdraw the tank downward.

7 It is advisable to remove the main filter whilst the tank is detached from the frame by unscrewing the large hexagon which is part of the main feed union. Wash the filter in petrol until it is clean and let it dry. Before replacing it make sure the sealing washer is in good condition and give the tank itself a thorough wash out with petrol. Tighten the hexagon fully, to prevent oil leaks.

8 Replace the tank by reversing the dismantling procedure. When the flexible oil pipes are replaced, check that the inside bore is not rough or beginning to flake. Under these circumstances, particles of rubber can break away and impede the flow of oil; renewal of the pipe is essential. The clips must be tightened fully, otherwise a pipe may come adrift and spray the rear tyre with oil.

9 Do not omit to replace the drain plug before the tank is refilled with engine oil of the recommended grade.

15 Speedometer head and tachometer head - removal and replacement

1 The speedometer and tachometer heads are each secured to a bracket that bolts to the fork top yoke. The bracket is rubber-mounted to damp out vibration; both instruments are held to the bracket by studs which project from the base of the casing. On late models the instruments are mounted within rubber cups, attached to brackets that fit under each top fork nut.

2 To remove either instrument, detach the drive cable by unscrewing the gland nut where the cable enters the instrument body. Pull out the bulb holder complete with the bulb used to illuminate the dial and unscrew the two nuts that secure the instrument to the mounting bracket, taking care not to displace the shakeproof washers. The instrument can now be lifted away.

3 Apart from defects in either the drive or the drive cable, a speedometer or tachometer that malfunctions is difficult to repair. Fit a replacement, or alternatively entrust the repair to an instrument repair specialist, bearing in mind that an efficient speedometer is a statutory requirement. If, in the case of a speedometer, the mileage recordings also cease, it is highly probable that either the drive cable or the drive is at fault and not the speedometer head itself. It is very rare for all recordings to fail simultaneously.

16 Speedometer and tachometer drive cables - examination and renovation

1 It is advisable to detach the speedometer and tachometer drive cables from time to time, in order to check whether they are adequately lubricated and whether the outer covers are compressed or damaged at any point along their run. A jerky or sluggish movement at the instrument head can often be attributed to a cable fault.

2 To grease the cable, uncouple both ends and withdraw the inner cable. After removing the old grease, clean with a petrol soaked rag and examine the cable for broken strands or other damage.

3 Regrease the cable with high melting point grease, taking care not to grease the last six inches closest to the instrument head. If this precaution is not observed, grease will work into the instrument and immobilise the sensitive movement.

4 If the cable breaks, it is usually possible to renew the inner cable alone, provided the outer cable is not damaged or compressed at any point along its run. Before inserting the new inner cable, it should be greased in accordance with the instructions given in the preceding paragraph. Try and avoid tight bends in the run of a cable because this will accelerate wear and make the instrument movement sluggish.

17 Dual seat - removal

1 The dual seat hinges on the left hand side of the machine when the retaining catch on the right hand side, below the dual seat, is released. A restraining wire limits the amount of travel.

2 To remove the dual seat, detach the restraining wire from the seat underpan where it is retained by a self-tapping screw. If either the forward or rearward hinge is removed, by unscrewing the two bolts that secure it to the tapped inserts in the seat underpan, the seat can be slid off the remaining hinge pivot and lifted away.

18 Left hand side cover - removal

1 The left hand side cover, which contains the tool roll, is held in position by a panel knob having a large milled head. If the knob is unscrewed completely, the cover can be drawn off the two projections from the rear subframe and lifted away.

2 Do not lose the two grommets which fit over the frame tube projections.

106

FIG. 6.6. SWINGING ARM FORK AND REAR SUSPENSION UNITS

No.	Description	No.	Description	No.	Description
1	Swinging arm fork	13	Spring washer - 2 off	25	Nut - 4 off
2	Pivot bush - 2 off	14	Nut - 2 off	26	Chainguard
3	Grease nipple	15	Rear suspension unit - 2 off	27	Grommet
4	Fibre washer	16	Bonded bush - 4 off	28	Bolt
5	Sleeve bush - 2 off	17	Spring retaining collets - 4 off	29	Rubber tube for chain oiler
6	Distance piece	18	Outer shroud - 2 off	30	Plain washer
7	Flanged washer - 2 off	19	Inner shroud - 2 off	31	Self locking nut
8	Pivot pin	20	Spring, 145 lbs rating - 2 off	32	Stepped bolt
9	Tab washer	21	Fixing bolt - 4 off	33	Plain washer
10	Nut	22	Plain washer - 2 off	34	Nut
11	Rear brake torque stay	23	Spring washer - 2 off	35	Clip
12	Bolt	24	Plain washer - 3 off	36	Nut
				37	Bolt
				38	Spring washer
				39	Brake operating rod
				40	Split pin
				41	Adjuster nut
				42	Stop lamp switch
				43	Screw - 3 off
				44	Self locking nut - 5 off
				45	Switch lever
				46	'D' washer
				47	Spring
				48	'O' ring - 2 off

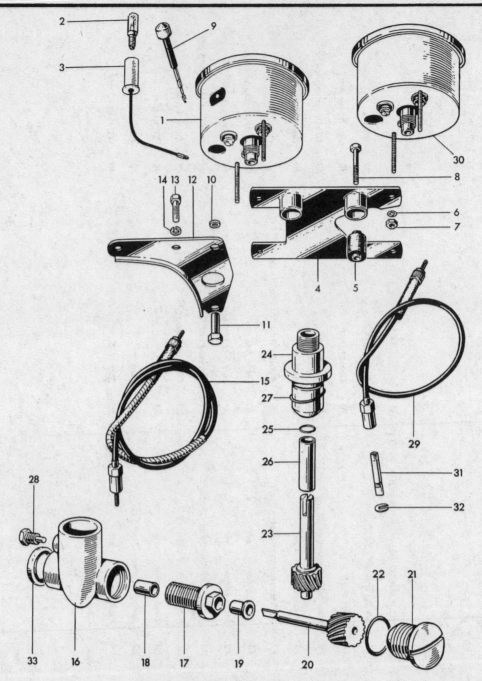

FIG. 6.7. SPEEDOMETER AND TACHOMETER

1	Speedometer head	9	Trip control*	17	Securing screw	25	'O' ring	33	Sealing washer
2	Bulb, 3W M.E.S.	10	Distance piece - 2 off*	18	Bush	26	Bush		* TR6 and T120's
3	Bulb holder	11	Sleeve nut - 2 off*	19	Flanged bush	27	'O' ring		
4	Mounting bracket	12	Bracket*	20	Driving gear	28	Set screw		
5	Metalastik bush - 2 off	13	Bolt - 2 off	21	End cap	29	Tachometer cable (28 inches long)		
6	Washer - 2 off	14	Spring washer - 2 off*	22	'O' ring	30	Tachometer head		
7	Nut - 2 off	15	Speedometer cable (65 inches long)	23	Driven gear	31	Spade		
8	Bolt - 2 off	16	Tachometer gearbox	24	Driven gear housing	32	Circlip		

19 Petrol tank embellishments - removal

1 Some models have metal tank badges secured one to each side of the tank. Each is held in position by two screws which, when released, permits removal of the badge.

2 There is also a chromium plated strip down the middle of the tank that hides the otherwise unsightly seam. The strip hooks around the nose of the tank at the front and is retained at the rear by a crosshead bolt which passes through the rear most tank mounting lug, and is secured by a nut.

3 In accordance with well established Triumph practice, a great number of machines are fitted with a tank top luggage carrier which takes the form of a chromium plated grille. The grille is held to the tank top by four slotted head screws which mate up with tapped inserts in the tank top.

4 Rubber knee grips preserve the paintwork on the sides of the petrol tank. They are retained in place by an adhesive compound.

20 Steering head lock

1 A steering head lock is built into the upper fork yoke which will lock the handlebars on full left lock when it is operated. The lock is held in position by a grub screw blanked off with a sealing washer. In operation, a tongue from the lock projects through a plate attached to the steering head when aligned correctly.

2 If the lock is changed, the key must also be changed, since the locks are numbered individually.

21 Fairing attachment points

Some models are fitted with a cast-in lug at the top of the steering head to facilitate the fixing of a fairing. Two rod-like attachments pass through the base of the steering head, the lower extremities of which form the lock stops for the forks.

22 Sidecar alignment

1 Using conventional fittings, little difficulty is experienced when attaching a sidecar to any of the Triumph 650/750 cc unit-construction twins.

2 Good handling characteristics of the outfit will depend on the accuracy with which the sidecar is aligned. Provided the toe-in and lean-out are within prescribed limits, good handling characteristics should result, leaving scope for other minor adjustments about which opinions vary quite widely.

3 To set the toe-in, check that the front and rear wheels of the motor cycle are correctly in line and adjust the sidecar fittings so that the sidecar wheel is approximately parallel to a line drawn between the front and rear wheels of the machine. Re-adjust the fittings so that the sidecar wheel has a slight toe-in toward the front wheel of the motor cycle, as shown in Fig 6.6. When the amount of toe-in is correct, the distance 'B' should be from 3/8 in to 3/4 in less than the distance at 'A'.

4 Lean-out is checked by attaching a plumb line to the handlebars and measuring the distance between 'C' and 'D' as shown in Fig 6.7. Lean-out is correct when the distance 'C' is approximately 1 inch greater than at 'D'.

23 Cleaning the machine - general

1 After removing all surface dirt with a rag or sponge washed frequently in clean water, the application of car polish or wax will give a good finish to the machine. The plated parts should require only a wipe over with a damp rag, followed by polishing with a dry rag. If, however, corrosion has taken place, which may occur when the roads are salted during the winter, a proprietary chrome cleaner can be used.

2 The polished alloy parts will lose their sheen and oxidise slowly if they are not polished regularly. The sparing use of metal polish or a special polish such as Solvol Autosol will restore the original finish with only a few minutes labour.

3 The machine should be wiped over immediately after it has been used in the wet so that it is not garaged under damp conditions which will encourage rusting and corrosion. Make sure to wipe the chain and if necessary re-oil it to prevent water from entering the rollers and causing harshness with an accompanying rapid rate of wear. Remember there is little chance of water entering the control cables if they are lubricated regularly, as recommended in the Routine maintenance section.

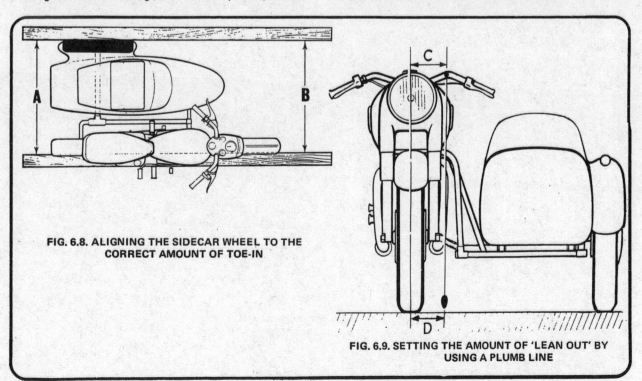

FIG. 6.8. ALIGNING THE SIDECAR WHEEL TO THE CORRECT AMOUNT OF TOE-IN

FIG. 6.9. SETTING THE AMOUNT OF 'LEAN OUT' BY USING A PLUMB LINE

24 Fault diagnosis

Symptom	Reason/s	Remedy
Machine is unduly sensitive to road surface irregularities	Fork and/or rear suspension units damping ineffective	Check oil level in forks. Renew suspension units.
Machine rolls at low speeds	Steering head bearings overtight or damaged	Slacken bearing adjustment. If no improvement, dismantle and inspect head races.
Machine tends to wander. Steering imprecise	Worn swinging arm suspension bearings	Check and if necessary renew bushes.
Fork action stiff	Fork legs twisted in yokes or bent	Slacken off wheel spindle clamps, yoke pinch bolts and fork top nuts. Pump forks several times before retightening from bottom. Straighten or renew bent forks.
Forks judder when front brake is applied	Worn fork bushes Steering head bearings slack	Strip forks and renew bushes. Re-adjust to take up play.
Wheels seem out of alignment	Frame distorted through accident damage	Check frame after stripping out. If bent, specialist repair or renewal is necessary.
Machine handles badly under all types of condition	General frame and fork distortion as result of a previous accident or broken frame tube	Strip frame and check for alignment very carefully. If wheel track is out, renew or straighten parts involved. Broken tube will be self-evident.

Chapter 7 Wheels, brakes and final drive

Contents

Specifications

Wheels

Rim sizes -	front	WM2-19
	rear	WM2-18 or WM3-18 *
Tyre sizes -	front	3.25 x 19
	rear	3.50 x 18 or 4.00 x 18 *

   * Export and post 1970 models

Brakes

Type -	front	Twin leading shoe (drum) or disc *
	rear	Single leading shoe (drum)
Diameter -	front	8 inch drum
*	front	10 inch disc
	rear	7 inch drum

Final drive chain

Size	5/8 in x 3/8 in
Number of pitches	106 (solo)

Rear wheel sprocket

Number of teeth	46 or 47 *

   * Post 1970 models

Tyre pressures

All models 1963-67	20 psi front, 18 psi rear
All other models	24 psi front, 25 psi rear

If a pillion passenger is carried increase the front tyre pressure by 4 psi and the rear tyre pressure by 6 psi.
Above figures recommended for a 12 stone solo rider.

1 General description

The unit-construction twin cylinder models are fitted with a 19 inch diameter front wheel and an 18 inch diameter rear wheel. All models have a 3.25 inch section tyre fitted to the front wheel; the rear wheel has either a 3.50 inch section tyre, or in the case of the post 1970 models, a tyre of 4.00 inch section. The American export models, irrespective of the year of manufacture, are fitted as standard with the larger section rear tyre.

Until 1971 it was customary to fit an 8 inch diameter internal expanding brake of the twin leading shoe to the front wheel and a 7 inch diameter internal expanding brake of conventional design to the rear wheel. Thereafter an 8 inch diameter front brake of the twin leading shoe type, with its adjustment modelled on car practice, but cable operated, was substituted for the original design, whilst the rear brake remained virtually unchanged. These models are identified by the conical shape of the hubs.

Both of the 750 cc models have an hydraulic disc brake

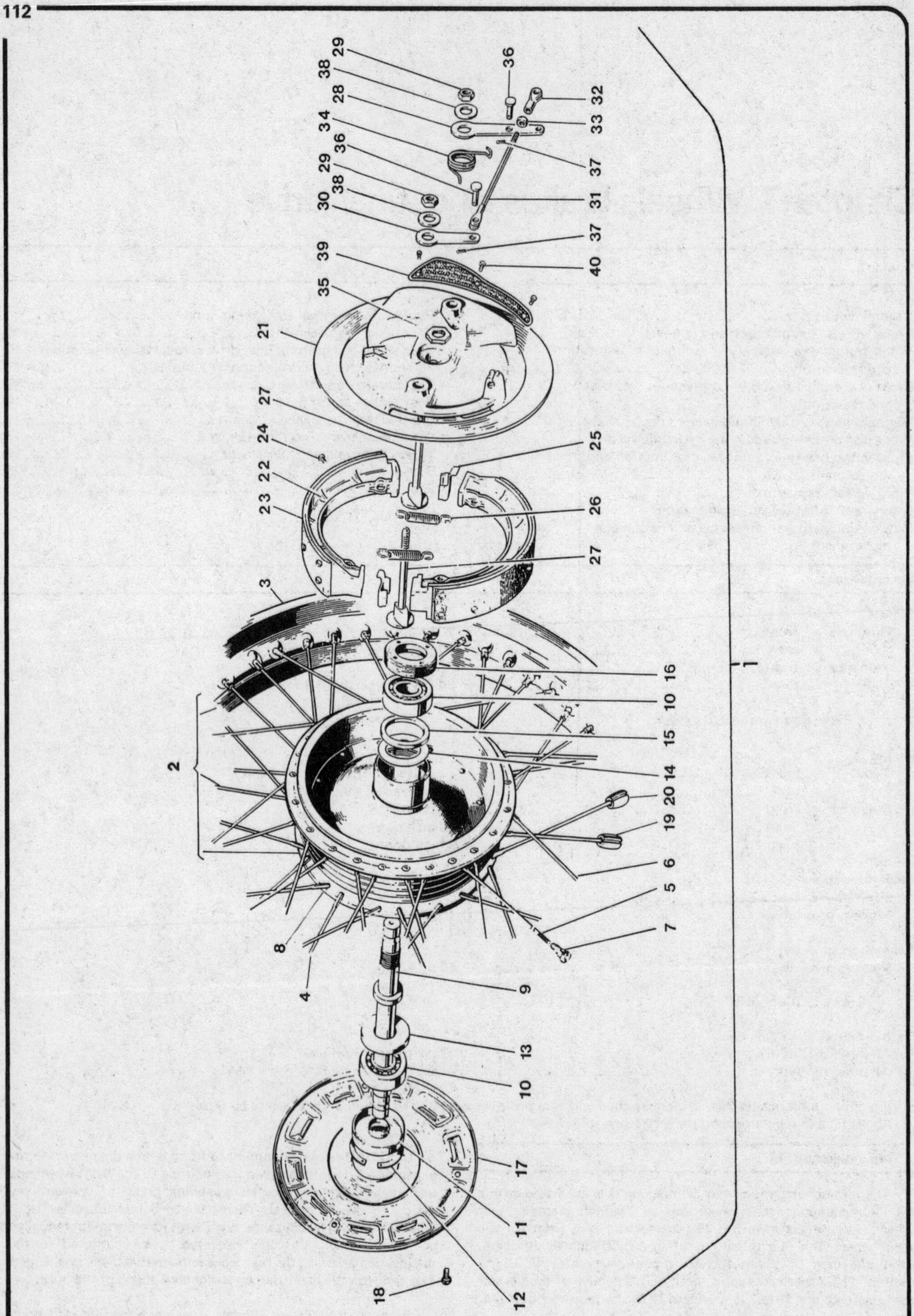

1
2
3
4
5
6
7
8
9
10
11
12
13
14
15
16
17
18
19
20
21
22
23
24
25
26
27
28
29
30
31
32
33
34
35
36
37
38
39
40

FIG. 7.1. FRONT WHEEL ASSEMBLY (EARLY TYPE)

1 Front wheel complete
2 Rim, spokes and hub assembly
3 Front wheel rim (WM2 - 19)
4 Spoke (5 5/8 inch straight) complete with nipple - 20 off
5 Spoke (4 25/32 inch 78º) complete with nipple - 10 off
6 Spoke (4 7/8 inch 100º) complete with nipple - 10 off
7 Spoke nipple - 40 off
8 Hub and brake drum
9 Wheel spindle
10 Bearing - 2 off
11 Dust cover
12 Circlip
13 Grease retainer
14 Grease retainer (brake side)
15 Backing ring
16 Retaining ring
17 Cover plate
18 Fixing screw - 3 off
19 Balance weight - ½ oz
20 Balance weight - 1 oz

21 Brake anchor plate
22 Brake shoe complete with lining - 2 off
23 Brake lining - 2 off
24 Brass rivet - 16 off
25 Abutment pad - 2 off
26 Shoe return spring - 2 off
27 Brake cam - 2 off
28 Brake cam lever (front)
29 Nut - 2 off
30 Brake cam lever (rear)
31 Rod assembly
32 Fork end, threaded
33 Locknut
34 Lever return spring
35 Anchor plate nut
36 Clevis pin - 2 off
37 Split pin - 2 off
38 Plain washer - 2 off
39 Anchor plate gauze (for air scoop)
40 Taptite screw

114

1

2 3 4 5 6 7 8 9 10 11 12 13 14 15 16 17 18 19 20 21 22 23 24 25 26 27 28 29 30 31 32 33 34 35

FIG. 7.2. FRONT WHEEL ASSEMBLY (CONICAL HUB)

1 Front wheel complete
2 Rim, spokes and hub assembly
3 Wheel rim (WM2 - 19)
4 Brake drum and hub
5 Spoke (right hand, inner) - 10 off
6 Spoke (right hand, outer) - 10 off
7 Spoke (left hand) - 40 off
8 Nipple - 40 off
9 Balance weight - short
10 Balance weight - long
11 Bearing - 2 off
12 Circlip, right-hand
13 Grease retainer, right hand
14 Lock ring, right hand
15 Grease cap, inner left hand
16 Grease cap, outer left hand
17 Circlip, left hand
18 Distance piece

19 Wheel spindle
20 Wheel spindle nut
21 Brake drum grommet
22 Nut for brake anchor
23 Washer for brake anchor
24 Brake anchor plate
25 Brake lever (front)
26 Brake lever (rear)
27 'O' ring for brake cam - 2 off
28 Tappet, brake cam - 2 off
29 'O' ring, brake tappet
30 Brake shoe complete - 2 off
31 Return spring - 2 off
32 Cage, brake shoe adjuster - 2 off
33 Micram adjuster - 2 off
34 Abutment pressing - 2 off
35 Anchor plate stud

fitted to the front wheel to match the extra performance.

The rear wheel is of the quickly detachable variety which means the wheel can be removed, if desired, leaving the final drive chain and sprocket in position. This arrangement does not apply to later models fitted with conical hubs, or to the 750 cc models.

2 Front wheel - examination and removal

1 Place the machine on the centre stand so that the front wheel is raised clear of the ground. Spin the wheel and check for rim alignment. Small irregularities can be corrected by tightening the spokes in the affected area, although a certain amount of skill is necessary if over-correction is to be avoided. Any 'flats' in the wheel rim should be evident at the same time. These are more difficult to remove with any success and in most cases the wheel will have to be rebuilt on a new rim. Apart from the effect on stability, there is greater risk of damage to the tyre bead and walls if the machine is run with a deformed wheel, especially at high speeds.

2 Check for loose or broken spokes. Tapping the spokes is the best guide to the correctness of tension. A loose spoke will produce a quite different note and should be tightened by turning the nipple in an anticlockwise direction. Always check for run-out by spinning the wheel again.

3 If several spokes require retensioning or there is one that is particularly loose, it is advisable to remove the tyre and tube so that the end of each spoke that projects through the nipple after retensioning can be ground off. If this precaution is not taken, the portion of the spoke that projects may chafe the inner tube and cause a puncture.

3 Front drum brake assembly - examination, renovation and reassembly

1 The front brake assembly complete with brake plate can be withdrawn from the front wheel by following the procedure in Chapter 6, section 2, paragraphs 1 to 3. In the case of the post 1970 models, fitted with conical hubs, a slightly different procedure is necessary when detaching the brake cable from the rearmost brake operating arm. It is also necessary to slacken the torque lug nut on the inner portion of the right hand fork leg before the wheel can be freed.

2 An anchor plate nut retains the brake plate on the front wheel spindle. When this nut is removed, the brake plate can be drawn away, complete with the brake shoe assembly.

3 Examine the condition of the brake linings. If they are wearing thin or unevenly, the brake shoes should be relined or renewed.

4 To remove the brake shoes from the brake plate, pull them apart whilst lifting them upward, in the form of a V. When they are clear of the brake plate, the return springs can be removed and the shoes separated. Do not lose the abutment pads fitted to the leading edge of each shoe.

5 The brake linings are rivetted to the brake shoes and it is easy to remove the old linings by cutting away the soft metal rivets. If the correct Triumph replacements are purchased, the new linings will be supplied ready-drilled with the correct complement of rivets. Keep the lining tight against the shoe throughout the rivetting operation and make sure the rivets are countersunk well below the lining surface. If workshop facilities and experience suggest it would be preferable to obtain replacement shoes, ready lined, costs can be reduced by making use of the Triumph service exchange scheme, available through Triumph agents.

6 Before replacing the brake shoes, check that both brake operating cams are working smoothly and not binding in their pivots. The cams can be removed for cleaning and greasing by unscrewing the nut on each brake operating arm and drawing the arm off, after its position relative to the cam spindle has been marked so that it is replaced in exactly the same position. The spindle and cam can then be pressed out of the housing in the back of the brake plate.

7 Check the inner surface of the brake drum on which the brake shoes bear. The surface should be smooth and free from score marks or indentations, otherwise reduced braking efficiency is inevitable. Remove all traces of brake lining dust and wipe both the brake drum surface and the brake shoes with a clean rag soaked in petrol, to remove any traces of grease. Check that the brake shoes have chamfered ends to prevent pick-up or grab. Check that the brake shoe return springs are in good order and have not weakened.

8 To reassemble the brake shoes on the brake plate, fit the return springs first and force the shoes apart, holding them in a V formation. If they are now located with the operating cams they can usually be snapped into position by pressing downward. Do not use excessive force or the shoes may distort permanently. Make sure the abutment pads are not omitted.

9 A different type of brake unit is fitted to the post 1970 models which have wheels with conical hubs. Although the operating principle is the same, car-type brake shoe expanders are fitted. A micram adjuster is fitted in place of the abutment pads used previously, providing a means of compensating for brake lining wear without having to reduce the angle of the brake operating arms. The brake unit can be dismantled and reassembled by using the procedure already described.

4 Front disc brake assembly - examination, renovation and reassembly

1 The front wheel complete with the brake disc can be withdrawn from the front wheel fork after removing the two clamp bolts retaining each split clamp to the lower fork ends and withdrawing both clamps. The disc is secured to the left hand side of the wheel hub by four domed nuts which, when removed, will release the disc. It should not be necessary to remove the disc unless it is badly scored or requires renewal for some other reason.

2 When the wheel is removed, it is advisable to place a clean spacer, such as a piece of wood or metal, between the brake pads to prevent them being ejected if the front brake is unintentionally applied. This precaution is not necessary if the caliper piston assemblies are to receive attention.

3 The friction pads will lift out of the brake caliper if they are turned slightly. Inspect the friction faces for excessive wear, uneven wear or scoring. Renew both pads if there is any doubt about the condition of either one. Pads should always be renewed in pairs, never singly.

4 Clean out the recesses into which the pads fit and the exposed end of the pistons, using a soft brush. Do not on any account use a solvent cleanser or a wire brush. Finish off by giving the piston faces and the friction pad recesses just a smear of brake fluid.

5 If it is found that the pistons do not move freely or are seized in position, the caliper is in urgent need of attention and must be removed, drained and overhauled. Seek advice from a professional experienced with motor cycle disc brakes. If a piston is seized, the only satisfactory course of action is renewal of the complete brake caliper unit.

6 To remove the brake caliper unit from the machine, unscrew the union where the hydraulic fluid pipe enters the unit and drain the complete hydraulic system into a clean container. Never re-use brake fluid. The caliper unit can now be detached from its mounting on the lower left hand fork leg by removing the two retaining bolts.

7 If the caliper unit shows evidence of brake fluid leakage, accompanied by the need to top up the hydraulic fluid reservoir at regular intervals, the piston seals require renewal. This is a simple task which is carried out as follows: remove the front wheel complete with disc as described in Section 4.1. Lift both friction pads out of position and mark the friction faces so that they are replaced in an identical position. Drain the hydraulic system by placing a clean receptacle below the unit to catch the hydraulic fluid and squeezing the handlebar lever so that both pistons are expelled to release the fluid. Unscrew the caliper end plug which has two peg holes and will require the

use of the correct peg spanner tool because it is a tight fit. Remove the two pressure seals from their respective grooves, using a blunt nosed tool to ensure the grooves are not damaged in any way.

8   Wet the new seals with hydraulic fluid and insert the first seal into the innermost bore, making sure it has seated correctly. The diameter of the seal is larger than that of the groove into which it fits, so that a good interference fit is achieved. Furthermore, the sections of the seal and seal groove are different to ensure the sealing edge is proud of the groove. Wet one of the pistons with hydraulic fluid and insert it through the outer cylinder (uncovered by removal of the caliper end plug) so that it passes through into the innermost cylinder bore, closed end facing inward. Check that it enters the seal squarely and leave it protruding approximately 5/16 in (8 mm) from the mouth of the inner bore.

9   Fit the seal and piston in the outer bore of the caliper unit using an identical procedure. Fit a new O ring seal and replace the end plug, tightening it to a torque setting of 26 lb ft. Replace the friction pads in their original positions after checking that all traces of fluid used to lubricate the various components during assembly have been removed, replace the front wheel and refill the master cylinder reservoir with the correct grade of hydraulic fluid. It will be necessary to bleed the system before the correct brake action is restored by following the procedure described fully in Section 6.

10 Note that all these operations must be carried out under conditions of extreme cleanliness. The brake caliper unit must be cleaned thoroughly before dismantling takes place. If particles of grit or other foreign matter find their way into the hydraulic system there is every chance that they will score the precision made parts and render them inoperative, necessitating expensive replacements.

5 Master cylinder - examination and replacing seals

1   The master cylinder and hydraulic fluid reservoir take the form of a combined unit mounted on the right hand side of the handlebars, to which the front brake lever is attached. The master cylinder is actuated by the front brake lever and applies hydraulic pressure through the system to operate the front brake when the handlebar lever is manipulated. The master cylinder pressurises the hydraulic fluid in the pipe line which, being incompressible, causes the pistons to move within the brake caliper unit and apply the friction pads to the brake disc. It follows that if the piston seals of the master unit leak, hydraulic fluid will be lost and the braking action rendered much less effective.

2   Before the master unit can be removed and dismantled, the system must be drained. Place a clean container below the brake caliper unit and attach a plastic tube from the bleed screw of the caliper unit to the container. Open the bleed screw one complete turn and drain the system by operating the brake lever until the master cylinder reservoir is empty. Close the bleed screw and remove the pipe.

3   To gain access to the master cylinder, disconnect the front brake stop lamp switch by pulling off the spade terminal connections. Lift away the switch cover and detach the hose from the master cylinder by unscrewing the union joint. Remove the four screws securing the unit to the handlebars by means of a split clamp and withdraw the unit complete with integral reservoir.

4   Remove the reservoir cap and bellows seal from the top of the reservoir. Remove the front brake stop lamp switch by unscrewing it from the master cylinder body. Release the handlebar lever by withdrawing the pivot bolt. The rubber boot over the master cylinder piston is retained in position by a special circlip having ten projecting ears. If three or four adjacent ears are lifted progressively, the circlip can gradually be lifted away until it clears the mouth of the bore and is released completely. It will most probably come away with the piston together with the secondary cup.

3.1 Brake plate will lift away when front wheel is detached from forks

3.4 Pull and lift brake shoe upwards to separate from brake plate

5   Remove the primary cup washer, cup spreader and bleed valve assembly which will remain within the cylinder bore. They are best displaced by applying gentle air pressure to the hose union bore.

6   Examine the cylinder bore for wear in the form of score marks or surface blemishes. If there is any doubt about the condition of the bore, the master cylinder must be renewed. Check the brake operating lever for pivot bore wear, cracks or fractures, the hose union and switch threads, and the piston for signs of scuffing or wear. Finally, check the brake hose for cuts, cracks or other signs of deterioration.

7   Before replacing the component parts of the master cylinder, wash them all in clean hydraulic fluid and place them in order of assembly on a clean, dust-free surface. Do not wipe them with a fluffy rag; they should be allowed to drain. Particular attention should be given to the replacement primary and secondary cup washers, which must be soaked in hydraulic fluid for at least fifteen minutes to ensure they are supple. Occasional kneading will help in this respect. Clean hands are essential.

8   Commence assembly by placing the unlipped side of the hollow secondary cup over the ground crown of the piston and work it over the crown, then the piston body and shoulder, until it seats in the groove immediately below the piston. Extreme care is needed during the entire operation which must be performed by hand. If a tool of any kind is used, it is highly

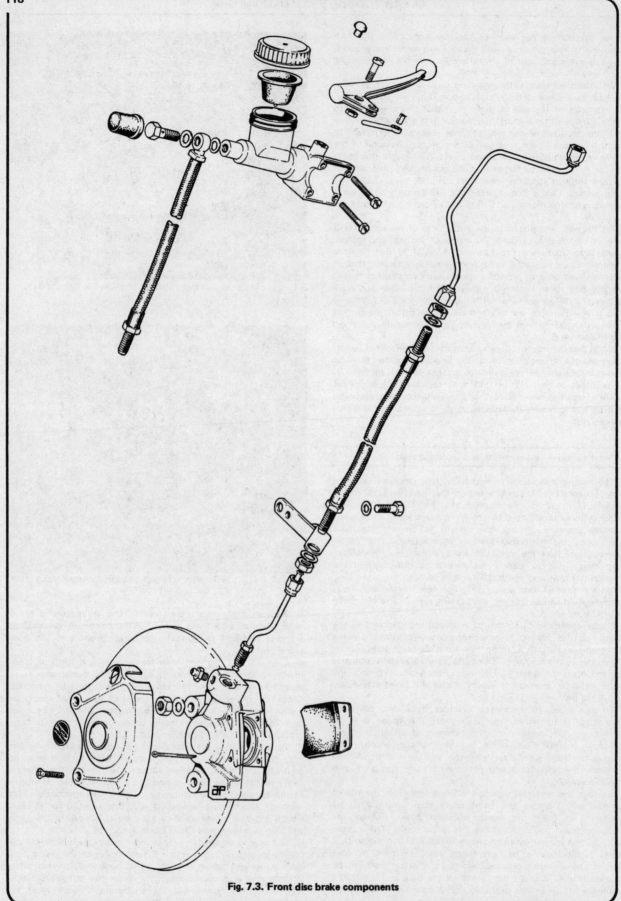

Fig. 7.3. Front disc brake components

probable that the lip seal will be damaged.

9 Fit the boot over the piston, open end toward the secondary cup, and engage the upper end in the piston groove, so that the boot seats squarely.

10 Assemble the trap valve spring over the plastic bobbin; the bobbin must seat squarely in the rubber valve base. Check that the small bleed hole is not obstructed, then insert the plastic spreader in the end of the trap valve spring furthest from the rubber valve. Offer the valve and spring assembly into the master cylinder bore, valve end first, keeping the cylinder bore upright. The casting can be held in a soft jawed vice during this operation to facilitate assembly, provided it is clamped very lightly.

11 Insert the primary cup into the bore with the belled-out end innermost. Insert the primary cup washer with the dished portion facing the open end of the bore. Insert the piston, crown end first, into the bore and locate the ten ear circlip over the boot, so that the set of the ears faces away from the cylinder. Apply pressure with a rotary motion and check that the lip of the secondary cup enters the cylinder bore without damaging the lip. Maintain pressure and engage the lower shoulder of the boot with the counter bore of the cylinder that acts as its seating. Work the boot retaining circlip into position, whilst still maintaining pressure on the piston assembly.

12 Still maintaining pressure on the piston assembly, feed the brake lever into position at the fulcrum slot and align the pivot holes so that the pivot bolt can be replaced and locked with the locknut.

13 It must be emphasised that reassembly of the master cylinder is a very tricky and delicate operation in which great care has to be exercised to ensure the replacement seals are not damaged. The assistance of another person during reassembly is advisable since without previous experience it is difficult to maintain the required pressure on the piston assembly whilst the final operations are completed.

14 Replace the master cylinder on the handlebars and refill the reservoir with hydraulic brake fluid. The system must now be bled.

6 Front disc brake - bleeding the hydraulic system

1 As mentioned earlier, brake action is impaired or even rendered inoperative if air is introduced into the hydraulic system. This can occur if the seals leak, the reservoir is allowed to run dry or if the system is drained prior to the dismantling of any component part of the system. Even when the system is refilled with hydraulic fluid air pockets will remain and because air will compress, the hydraulic action is lost.

2 Check the fluid content of the reservoir and fill almost to the top. Remember that hydraulic brake fluid is an excellent paint stripper, so beware of spillage, especially near the petrol tank.

3 Place a clean glass jar below the brake caliper unit and attach a clear plastic tube from the caliper bleed screw to the container. Place some clean hydraulic fluid in the container so that the pipe is always immersed below the surface of the fluid.

4 Unscrew the bleed screw one complete turn and pump the handlebar lever slowly. As the fluid is ejected from the bleed screw the level in the reservoir will fall. Take care that the level does not drop too low whilst the operation continues, otherwise air will re-enter the system, necessitating a fresh start.

5 Continue the pumping action with the lever until no further air bubbles emerge from the end of the plastic pipe. Hold the brake lever against the handlebars and tighten the caliper bleed screw. Remove the plastic tube AFTER the bleed screw is closed.

6 Check the brake action for sponginess, which usually denotes there is still air in the system. If the action is spongy, continue the bleeding operation in the same manner, until all traces of air are removed.

7 Bring the reservoir up to the correct level of fluid (within ½ inch of the top of the reservoir) and replace the bellows seal and cap. Check the entire system for leaks. Recheck the brake action.

8 Note that fluid from the container placed below the brake caliper unit whilst the system is bled should not be re-used.

7 Front wheel bearings - removal, examination and replacement

1 When the brake plate of machines fitted with a drum brake is removed, the bearing retainer within the brake drum will be exposed. This has a left hand thread and is removed by using either Triumph service tool Z76 or a centre punch. The bearing on the right hand side can then be displaced by striking the left hand end of the front wheel spindle, using the shoulder of the spindle to drive the bearing outward. The left hand bearing can be displaced in similar fashion, after the retaining circlip has been removed, if the wheel spindle is inserted from the other side. Take note of the various grease retainers and the circlip or backing ring, as appropriate, so that they are replaced correctly during reassembly.

2 Wash the bearings in a petrol/paraffin mix to remove all traces of old grease and oil. Clean out the hub and repack it with fresh high melting point grease.

3 When the bearings are dry, check them for play or signs of roughness when they are turned. If there is any doubt about their condition, renew them.

4 When fitting the bearings, first insert the inner left hand grease retainer. Pack the left hand bearing with grease and drive it into place with a drift of the correct diameter. Fit the outer dust cap, followed by the circlip, which must locate correctly with the retaining groove. Insert the wheel spindle so

7.1a Front wheel bearing retainer has left hand thread

7.1b Left hand bearing and dust cover are retained by circlip

that the shouldered end bears against the bearing from the inside of the hub and drive the bearing and grease retainer forward so that they are hard against the circlip holding them in position.

5 Withdraw the spindle and re-insert it the other way round. Refit the right hand grease retainer, repack the right hand bearing with grease and drive it into position. Screw in the retainer ring, remembering it has a left hand thread, and tighten it fully. Tap the spindle from the left hand end to centralise.

6 On disc brake hubs the procedure is essentially the same, noting that the hub components are reversed; ie the left-hand (disc) bearing is now secured by a threaded retainer. Note that the service tool required is 61-3694 and that the retainer has a normal right-hand thread, also that the spindle fixing nut must be unscrewed before the retainer can be disturbed.

8 Front wheel - replacement

1 Place the front brake assembly in the brake drum and align the front wheel so that the torque anchorage locates with the peg or stud on the lower right hand fork leg. This is most important because the anchorage of the front brake plate is dependent solely on the correct location of these parts.

2 Whilst holding the wheel in position, replace the split clamps which secure the wheel spindle to each fork end. Each clamp has two bolts or four nuts which must be tightened evenly and fully. Note that each end of the spindle has a raised groove which must locate with the clamp bolts on early models. Tighten the nut that secures the brake plate torque arm (conical hub models).

3 Reconnect the front brake cable and check that the brake functions correctly, especially if the adjustment has been altered or the brake operating arms have been removed and replaced during the dismantling operation. Recheck the tightness of the bolts in the split clamps.

9 Rear wheel - removal and examination

1 Place the machine on the centre stand and before removing the wheel, check for rim alignment, loose or broken spokes and other wheel defects by following the procedure applying to the front wheel, as described in Section 2.

2 Two types of rear wheel have been fitted, the standard or the quickly detachable type. The latter has the advantage of simplified removal, leaving the final drive chain and sprocket in position. The option is no longer available with the latest conical hub wheel which is not made in the quickly detachable version.

3 If the wheel is of the standard type, commence by disconnecting the final drive chain at the detachable spring link. The task is made easier if the link is first positioned so that it is on the rear wheel sprocket. Unwind the chain off the rear sprocket and lay it on a clean surface.

4 Take off the brake rod adjuster and pull the brake rod clear of the brake operating arm. Disconnect the torque stay by removing the nut where the stay joins the rear brake plate. If the stop lamp stays in the 'on' position, disconnect the snap connector in the lead.

5 Slacken both rear wheel spindle nuts and raise the rear chainguard by slackening the bottom nut of the left hand suspension unit. Remove the speedometer cable from the gearbox on the right hand side of the rear hub (if fitted) by unscrewing the gland nut and withdrawing the cable. Withdraw the wheel rearward until it drops from the frame ends complete with chain adjusters. It may be necessary to tilt the machine or raise it higher, so that there is sufficient clearance for the wheel to be taken away from the machine.

6 A different procedure is employed in the case of machines fitted with the quickly detachable rear wheel. It is necessary only to unscrew the gland nut from the speedometer drive gearbox (if fitted) and withdraw the cable, then unscrew and remove the wheel spindle from the right hand side of the

machine. If the shouldered distance piece between the frame end and the hub is removed, the wheel can be pulled sideways to disengage it from the brake drum centre (splined fitting) before it is lifted away. Note there is a rubber ring seal over the splines which is compressed when the wheel is in position. This acts as a grit seal and must be maintained in a good condition.

7.1c Right hand bearing will drive out from left

8.1 Torque anchorage MUST engage with fork leg, as shown

9.6a Disconnect gland nut from speedometer drive to free cable

9.6b Distance piece must be displaced before wheel can be freed

9.6c Wheel can be removed without disturbing chain or brake drum

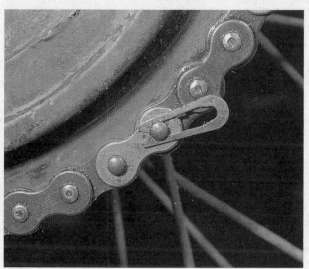

11.2a Detach final drive chain at rear sprocket

11.2b Remove nut from rear brake torque stay

11.2c Unscrew large nut around bearing sleeve

11.2d Note additional bearing within brake drum centre

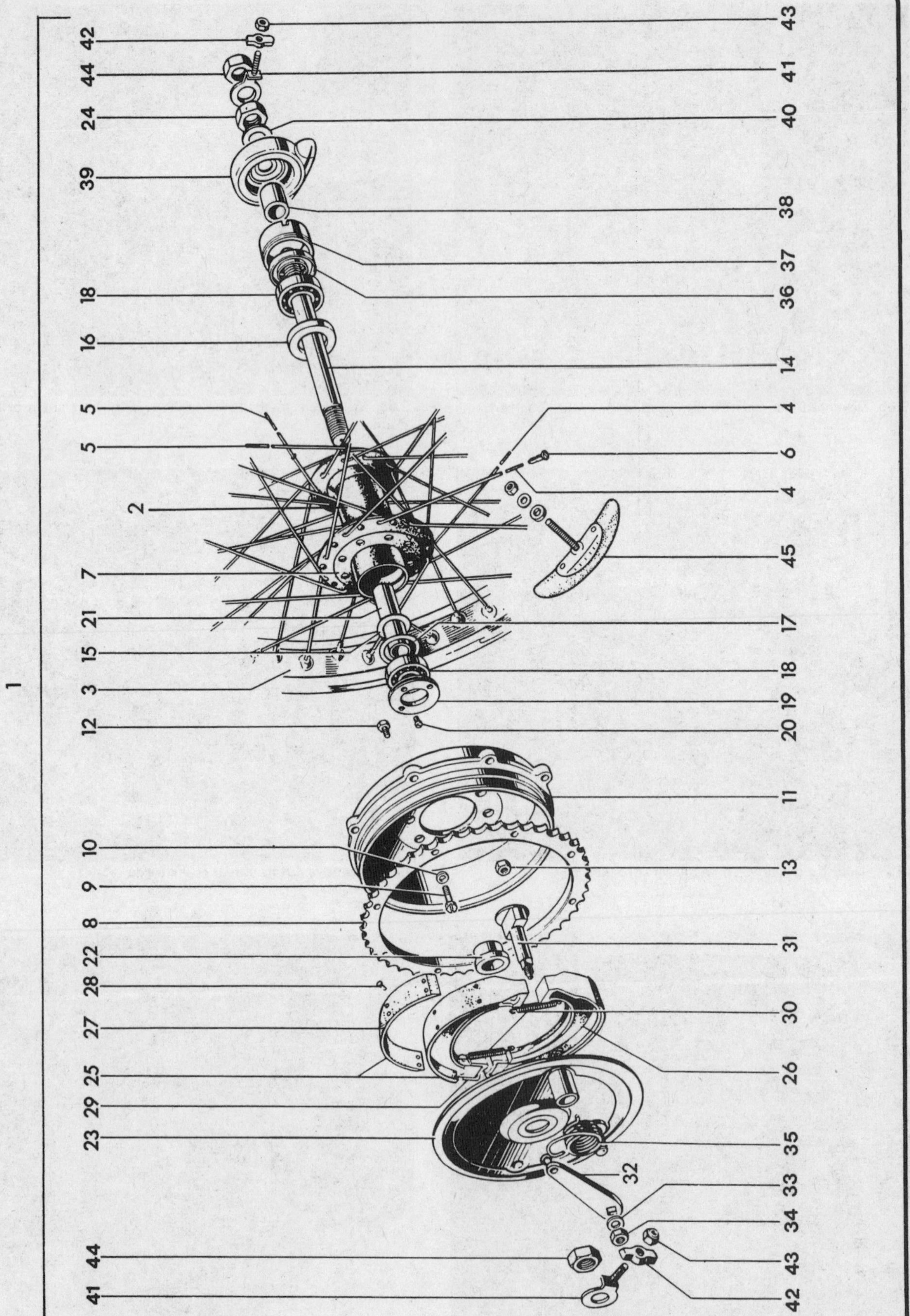
1
2
3
4
5
6
7
8
9
10
11
12
13
14
15
16
17
18
19
20
21
22
23
24
25
26
27
28
29
30
31
32
33
34
35
36
37
38
39
40
41
42
43
44
45

FIG. 7.4. REAR WHEEL ASSEMBLY (STANDARD TYPE)

1 Rear wheel complete
2 Rim, spokes and hub assembly
3 Rear wheel rim (WM2 - 18)
4 Spoke (7 9/16 inch 90°) complete with nipple - 20 off
5 Spoke (7 7/8 inch 90°) complete with nipple - 20 off
6 Spoke nipple - 40 off
7 Hub
8 Sprocket - 46 teeth
9 Bolt - 8 off
10 Serrated washer - 8 off
11 Brake drum
12 Bolt - 8 off
13 Self-locking nut - 8 off
14 Wheel spindle
15 Grease retainer (brake side)
16 Grease retainer
17 Backing ring
18 Bearing - 2 off
19 Retaining ring
20 Locking screw
21 Distance tube
22 Distance piece
23 Brake anchor plate

24 Nut
25 Leading brake shoe complete with lining
26 Trailing brake shoe complete with lining
27 Brake lining - 2 off
28 Brass rivet - 16 off
29 Thrust pad - 2 off
30 Brake shoe return spring - 2 off
31 Brake cam
32 Brake cam lever
33 Plain washer
34 Nut
35 Lever return spring
36 Grease retainer
37 Speedometer adaptor
38 Distance piece
39 Speedometer drive gearbox
40 Plain washer
41 Chain adjuster - 2 off
42 Adjuster end plate
43 Self-locking nut - 2 off
44 Spindle nut - 3 off
45 Tyre security bolt (WM 2)

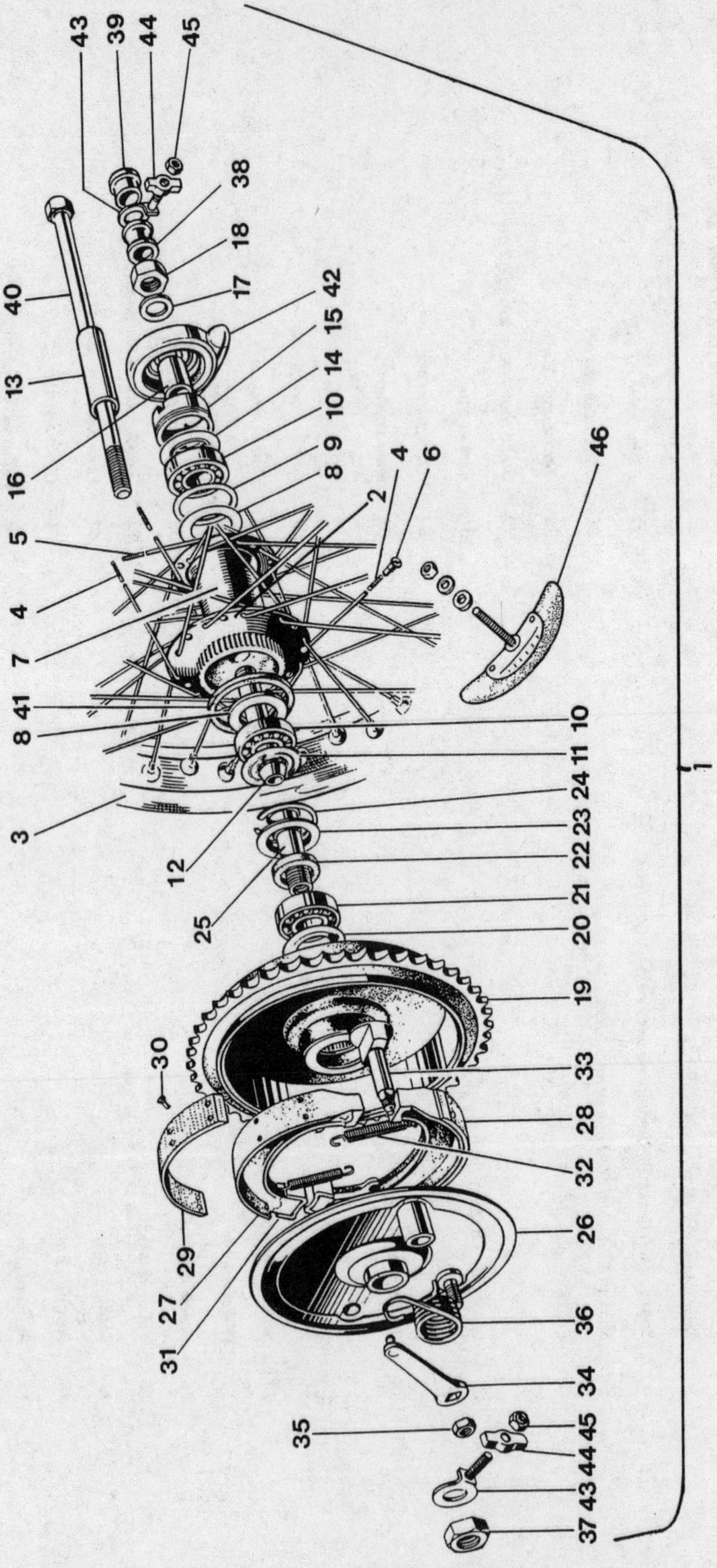
43
39
44
45
38
18
17
40
13
16
42
15
14
10
9
8
5
4
7
8
41
2
4
6
46
3
12
10
11
24
23
22
25
21
20
19
30
33
28
32
29
27
31
26
36
34
35
45
44
43
37
1

125

FIG. 7.5. REAR WHEEL ASSEMBLY (QUICKLY DETACHABLE TYPE)

1 Quickly-detachable wheel complete
2 Rim, spokes and hub assembly
3 Rear wheel rim (WM2 - 18)
4 Spoke (7 9/16 inch 90º) complete with nipple - 20 off
5 Spoke (7 7/8 inch 90º) complete with nipple - 20 off
6 Spoke nipple - 40 off
7 Hub
8 Grease retainer - 2 off
9 Backing ring
10 Bearing - 2 off
11 Dust cap
12 Bearing sleeve
13 Distance piece (inner)
14 Grease retainer
15 Locking ring
16 Distance piece (outer)
17 Plain washer
18 Nut
19 Sprocket and brake drum (46 teeth)
20 Grease retainer
21 Bearing
22 Felt washer
23 Bearing and felt retainer

24 Circlip
25 Bearing sleeve
26 Brake anchor plate
27 Leading brake shoe complete with lining
28 Trailing brake shoe complete with lining
29 Brake lining - 2 off
30 Brass rivet - 16 off
31 Thrust pad - 2 off
32 Brake shoe return spring - 2 off
33 Brake cam
34 Brake cam lever
35 Nut
36 Lever return spring
37 Nut
38 Inner distance piece
39 Outer distance piece
40 Spindle
41 Rubber sealing ring
42 Speedometer drive gearbox
43 Chain adjuster - 2 off
44 Adjuster end plate - 2 off
45 Self-locking nut - 2 off
46 Tyre security bolt (WM 2)

10 Rear wheel bearings - removal, examination and replacement

1 On standard rear wheels, unscrew the brake plate anchor nut and withdraw the brake plate and spacer, the wheel spindle and, where fitted; the speedometer drive gearbox and spacer. Remove the grub.screw which locks the left-hand bearing re-taining ring and use a peg spanner (service tool Z76) or a punch to unscrew the ring. **Note:** *check carefully whether the ring has a left- or right-hand thread before attempting to unscrew it and never use excessive force or the hub may be damaged.* On early models unscrew the right-hand bearing retaining nut, on later models unscrew the speedometer drive adaptor. **Note:** *check carefully whether the adaptor has a left- or right-hand thread before attempting to remove it and do not use excessive force or the hub may be damaged.* Insert a drift from the left-hand side and displace the central spacer so that a drift can be applied to the left-hand bearing inner race; it may be necessary to collapse the grease retainer. Tap out each bearing in turn, noting the location of the grease retainers, backing rings and dust caps which may be fitted. If damaged, the left-hand grease retainer can be gently hammered flat to restore its original shape.

2 On conical hubs, remove the spindle, brake backplate and the speedometer drive gearbox and spacer. Unscrew the left-hand bearing retaining ring (right-hand thread) using service tool 61-3694 or a punch and the speedometer drive adaptor (left-hand thread). Because the central spacer is shouldered and pressed into each bearing inner race a double-diameter drift must be fabricated which will fit closely inside the spacer and yet pass through the bearing inner race. Tap out either bearing first, complete with the spacer, then press the bearing off the spacer and tap out the remaining bearing. On reassembly, it is best to refit these bearings using a press in the reverse of the removal sequence.

3 On machines with quickly-detachable rear wheels, before engine number DU13375 (1963-'64), unscrew the two locknuts on the spindle sleeve right-hand end, push the sleeve out from the right and remove the small spacer from the right-hand bearing. Passing a drift through the hub from the opposite side tap out the inner races (and dust cap) followed by the outer races. On reassembly, reverse the removal procedure to refit the bearings, sleeve and spacer, then refit the locknuts. Tighten hard the inner locknut to settle the bearings then slacken it at least one flat (1/6 turn) and lock it by tightening the outer lock-nut securely. The sleeve and inner races should be able to rotate freely with no sign of free play.

4 On machines with quickly-detachable rear wheels from 1965-1970, ball journal bearings replace the previously-used taper roller type. Hold the bearing sleeve by the slot in its tapered end and unscrew the right-hand retaining nut. On 1965 models withdraw the spacer, on later models withdraw the washer, speedometer drive gearbox and its spacer; push the bearing sleeve out to the left. On 1965 models use service tool Z76 or a centre punch to unscrew the retaining ring (left-hand thread), on later models unscrew the speedometer drive adaptor; check first whether this has a left- or right-hand thread. Displace the central spacer as described in paragraph 1 above so that the right-hand bearing can be driven out, followed by the left-hand bearing. Note the location of the various grease retainers and backing rings to ensure correct reassembly.

5 On all models, check the bearings for wear as described in Section 7 and renew any worn or damaged components. Reassembly is the reverse of the dismantling operation; pack both bearings with grease as each is refitted.

11 Rear brake - removal and examination

1 If the rear wheel is of the standard type or if the rear wheel has been removed together with the brake drum and sprocket (quickly detachable wheels) the rear brake assembly is accessible when the brake plate is lifted away from the brake drum.

2 If the quickly detachable wheel has been removed in the recommended way (as described in Section 9.6) it will be necessary to detach the brake drum and sprocket assembly from the frame. This is accomplished by removing the final drive chain by detaching the spring link, unscrewing and removing the nut from the brake plate torque stay so that the latter can be pulled away, and unscrewing and removing the large nut around the bearing sleeve that supports the brake and sprocket assembly. Note that the brake drum has a bearing in its centre which should be knocked out, cleaned and examined before replacement.

3 The rear brake assembly is similar to that of the front wheel drum brake apart from the fact that it is of the single leading shoe type and therefore has only one operating arm. Use an identical procedure for examining and renovating the brake assembly to that described in Section 3, paragraphs 3 to 7. Note that the brake shoes are fitted with thrust pads and not abutment pads in this instance.

12 Rear brake - replacement

1 Reverse the dismantling procedure when replacing the front brake to give the correct sequence of operations. In the case of the quickly detachable rear wheel, the brake drum and sprocket should be fitted to the frame first to aid assembly.

2 It is important not to omit the rubber sealing ring of the quickly detachable wheel which fits over the splines of the hub and is compressed when the splines are located with those cut within the brake drum centre. The seal must be in good condition if it is to prevent the entry of road grit and other foreign matter which may cause rapid wear of the splines.

13 Rear wheel and gearbox final drive sprockets - examination

1 Before replacing the rear wheel, it is advisable to examine the rear wheel sprocket. A badly worn sprocket will greatly accelerate the rate of wear of the final drive chain and in an extreme case, will even permit the chain to ride over the teeth when the initial drive is taken up. Wear will be self-evident in the form of shallow or hooked teeth, indicating the need for early renewal.

2 In the case of the standard wheel, the sprocket is secured to the brake drum by eight nuts and bolts. When a quickly detachable wheel is fitted, the sprocket is an integral part of the brake drum, in which case the complete unit must be renewed.

3 The gearbox sprocket should also be inspected closely at the same time because it is considered bad practice to renew the one sprocket alone. A certain amount of dismantling work is necessary before the gearbox sprocket can be removed; Chapter 2, sections 4, 6 and 12.2 provide the relevant information.

14 Final drive chain - examination and lubrication

1 The final drive chain is not fully enclosed. The only lubrication provided takes the form of an oil bleed from the return pipe of the oil tank that distributes excess oil to the lower chain run.

2 Chain adjustment is correct when there is approximately ¾ inch play in the middle of the chain run, measured at either the top or the bottom. Always check at the tightest spot of the chain run with the rider seated normally.

3 If the chain is too slack, adjust by slackening the wheel spindle and/or wheel nuts and the nut of the torque arm stay, then drawing the wheel rearward by the chain adjusters at the end of each swinging arm fork. It is important to ensure that each adjuster is turned an equal amount so that the rear wheel is kept in alignment. When the correct adjusting point has been reached, push the wheel forward and tighten the wheel nuts and/or spindle, not forgetting the torque arm nut. Recheck the chain tension and wheel alignment, before the final tightening.

4 To check whether the chain needs renewing, lay it lengthwise in a straight line and compress it endwise so that all play is taken up. Anchor one end firmly, then pull endwise in the opposite direction and measure the amount of stretch. If it exceeds ¼ inch

Tyre changing sequence - tubed tyres

A Deflate tyre. After pushing tyre beads away from rim flanges push tyre bead into well of rim at point opposite valve. Insert tyre lever adjacent to valve and work bead over edge of rim.

B Use two levers to work bead over edge of rim. Note use of rim protectors

C Remove inner tube from tyre

When first bead is clear, remove tyre as shown

D

E When fitting, partially inflate inner tube and insert in tyre

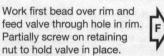

Work first bead over rim and feed valve through hole in rim. Partially screw on retaining nut to hold valve in place.

F

G Check that inner tube is positioned correctly and work second bead over rim using tyre levers. Start at a point opposite valve.

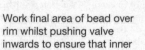

Work final area of bead over rim whilst pushing valve inwards to ensure that inner tube is not trapped

H

per foot, renewal is necessary. Never use an old or worn chain when new sprockets are fitted; it is advisable to renew the chain at the same time so that all new parts run together.

5 Every 2000 miles remove the chain and clean it thoroughly in a bath of paraffin before immersing it in a special chain lubricant such as Linklyfe or Chainguard. These latter types of lubricant are applied in the molten state (the chain is immersed) and therefore achieve much better penetration of the chain links and rollers. Furthermore, the lubricant is less likely to be thrown off when the chain is in motion.

6 When replacing the chain, make sure the spring link is positioned correctly, with the closed end facing the direction of travel. Replacement is made easier if the ends of the chain are pressed into the teeth of the rear wheel sprocket whilst the connecting link is inserted, or a simple 'chain-joiner' is used.

15 Front brake - adjustment

1 Brake adjustment is effected by the cable adjuster built into the front brake lever, which should be screwed outward to take up slack which develops in the operating cable as the brake shoes wear. Although adjustment is a matter of personal setting, there should never be sufficient slack in the cable to permit the lever to touch the handlebars before the brake is applied fully.

2 Eventually, braking action will be lost because cable adjustment has resulted in a poor angle between the brake operating arms and the direction of pull, causing loss of leverage. Provided the brake shoes are not badly worn, this can be corrected by slackening the adjuster fully and placing the machine on the centre stand. Remove the rubber plug from the front brake plate and insert a screwdriver so that each micram adjuster can be adjusted in turn. Start by turning the adjuster as far as it will go, so that the brake shoes are in contact with the brake drum. Back off two flats and check that the wheel is free to revolve. Turn the wheel half a revolution and repeat with the second adjuster. A further small adjustment may be necessary with the handlebar lever, to position the point of operation to the rider's liking. Do not omit to replace both rubber plugs in the brake plate and to recheck the brake before the machine is used on the road. This additional method of adjustment applies to machines fitted with conical hubs ONLY.

3 Front disc brakes require no adjustment; they are self-compensating.

16 Rear brake - adjustment

1 Rear brake adjustment is effected solely by the screwed adjuster at the extreme end of the brake operating rod. It should be screwed inward, to decrease the amount of play at the brake pedal. Always check after making adjustments to ensure that the brake shoes are not binding.

2 Brake adjustment will be necessary when slack in the rear chain is taken up. Because this involves moving the rear wheel backward in the frame, the rear brake adjuster may have to be slackened off a little.

3 After the rear brake has been adjusted, check the stop light action. It may be necessary to re-adjust the point at which the bulb lights by repositioning the clamp around the brake operating rod connected to the operating spring.

17 Front wheel - balancing

1 It is customary, on all high performance machines, to balance the front wheel complete with tyre and tube. The out of balance forces which exist are then eliminated and the handling of the machine improved. A wheel, which is badly out of balance produces throughout the steering, a most unpleasant hammering effect at high speeds.

2 One ounce and half ounce balance weights are available which can be slipped over the spokes and engaged with the square section of the spoke nipples. The balance weights are normally positioned diametrically opposite the tyre valve, which is usually

responsible for the out of balance factor.

3 When the wheel is spun it will come to rest with the heaviest point downward; balance weights should be added opposite to this point. Add or subtract balance weights until the wheel will rest in ANY position after it has been spun.

4 If balance weights are not available, wire solder wrapped around the spokes, close to the nipples, is an excellent substitute.

5 There is no necessity to balance the rear wheel for normal road use.

18 Speedometer drive gearbox - general

1 Models that do not have the speedometer drive taken from the gearbox require an external drive, which is usually taken from a speedometer drive gearbox fitted to the right hand side of the rear wheel. The drive is transmitted from the hub by means of a slotted locking ring which threads into the hub and performs the dual function of retaining the right hand wheel bearing.

2 Provided this gearbox is greased at regular intervals, it is unlikely to require attention during the normal life of the machine.

3 Speedometer drive gearboxes are not necessarily interchangeable, even though they may look similar. If a replacement has to be made, it is advisable to check the specification. The drive ratio is related to the size of the rear wheel and the section of tyre fitted, two variables which will have a marked effect on the accuracy of the speedometer reading.

19 Tyres - removal and replacement

1 At some time or other the need will arise to remove and replace the tyres, either as the result of a puncture or because a renewal is required to offset wear. To the inexperienced, tyre changing represents a formidable task yet if a few simple rules are observed and the technique learned the whole operation is surprisingly simple.

2 To remove the tyre from either wheel, first detach the wheel from the machine by following the procedure given in this Chapter whether the front or the rear wheel is involved. Deflate the tyre by removing the valve insert and when it is fully deflated, push the bead of the tyre away from the wheel rim on both sides so that the bead enters the centre well of the rim. Remove the locking cap and push the tyre valve into the tyre.

3 Insert a tyre lever close to the valve and lever the edge of the tyre over the outside of the wheel rim. Very little force should be necessary; if resistance is encountered it is probably due to the fact that the tyre beads have not entered the well of the wheel rim all the way round the tyre.

4 Once the tyre has been edged over the wheel rim, it is easy to work around the wheel rim so that the tyre is completely free on one side. At this stage, the inner tube can be removed.

5 Working from the other side of the wheel, ease the other edge of the tyre over the outside of the wheel rim furthest away. Continue to work around the rim until the tyre is free completely from the rim.

6 If a puncture has necessitated the removal of the tyre, re-inflate the inner tube and immerse it in a bowl of water to trace the source of the leak. Mark its position and deflate the tube. Dry the tube and clean the area around the puncture with a petrol soaked rag. When the surface has dried, apply rubber solution and allow this to dry before removing the backing from a patch and applying the patch to the surface.

7 It is best to use a patch of the self-vulcanising type, which will form a very permanent repair. Note that it may be necessary to remove a protective covering from the top surface of the patch, after it has sealed in position. Inner tubes made from synthetic rubber may require a special type of patch and adhesive if a satisfactory bond is to be achieved.

8 Before replacing the tyre, check the inside to make sure that the agent which caused the puncture is not trapped. Check the

outside of the tyre, particularly the tread area, to make sure nothing is trapped that may cause a further puncture.

9 If the inner tube has been patched on a number of past occasions, or if there is a tear or large hole, it is preferable to discard it and fit a new tube. Sudden deflation may cause an accident, particularly if it occurs with the front wheel.

10 To replace the tyre, inflate the inner tube just sufficiently for it to assume a circular shape. Then push it into the tyre so that it is enclosed completely. Lay the tyre on the wheel at an angle and insert the valve through the rim tape and the hole in the wheel rim. Attach the locking cap on the first few threads, sufficient to hold the valve captive in its correct location.

11 Starting at the point furthest from the valve, push the tyre bead over the edge of the wheel rim until it is located in the central well. Continue to work around the tyre in this fashion until the whole of one side of the tyre is on the rim. It may be necessary to use a tyre lever during the final stages.

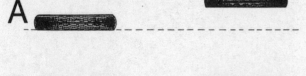

FIG.7.8. CHECKING WHEEL ALIGNMENT

A & C – Incorrect   B – Correct

12 Make sure there is no pull on the tyre valve and again commencing with the area furthest from the valve, ease the other bead of the tyre over the edge of the rim. Finish with the area close to the valve, pushing the valve up into the tyre until the locking cap touches the rim. This will ensure the inner tube

is not trapped when the last section of the bead is edged over the rim with a tyre lever.

13 Check that the inner tube is not trapped at any point. Reinflate the inner tube, and check that the tyre is seating correctly around the wall of the tyre on both sides, which should be equidistant from the wheel rim at all points. If the tyre is unevenly located on the rim, try bouncing the wheel when the tyre is at the recommended pressure. It is probable that one of the beads has not pulled clear of the centre well.

14 Always run the tyres at the recommended pressures and never under or over-inflate. See Specifications for recommended pressures.

15 Tyre replacement is aided by dusting the side walls, particularly in the vicinity of the beads, with a liberal coating of French chalk. Washing-up liquid can also be used to good effect, but this has the disadvantage of causing the inner surfaces of the wheel rim to rust.

16 Never replace the inner tube and tyre without the rim tape in position. If this precaution is overlooked there is good chance of the ends of the spoke nipples chafing the inner tube and causing a crop of punctures.

17 Never fit a tyre which has a damaged tread or side walls. Apart from the legal aspects, there is a very great risk of a blow-out, which can have serious consequences on any two-wheel vehicle.

18 Tyre valves rarely give trouble but it is always advisable to check whether the valve itself is leaking before removing the tyre. Do not forget to fit the dust cap which forms an effective second seal. This is especially important on a high performance machine, where centrifugal force can cause the valve insert to retract and the tyre to deflate without warning.

20 Security bolt

1 It is often considered necessary to fit a security bolt to the rear wheel of a high performance model because the initial take up of drive may cause the tyre to creep around the wheel rim and tear the valve from the inner tube. The security bolt retains the bead of the tyre to the wheel rim and prevents this occurrence.

2 A security bolt is fitted to the rear wheel of the Triumph 650/750 cc unit-construction twins as a safety precaution. Before attempting to remove or replace the tyre, the security bolt must be slackened off completely.

21 Fault diagnosis

Symptom	Reason/s	Remedy
Handlebars oscillate at low speeds	Buckle or flat in wheel rim, most probably front wheel	Check rim alignment by spinning wheel. Correct by retensioning spokes or rebuilding on new rim.
	Tyre not straight on rim	Check tyre alignment.
Machine lacks power and accelerates poorly	Brakes binding	Warm brake drum provides best evidence. Re-adjust brakes.
Brakes grab when applied gently	Ends of brake shoes not chamfered	Chamfer with file.
	Elliptical brake drum	Lightly skim in lathe (specialist attention required).
Front brake feels spongy	Air in hydraulic system (disc brake only)	Bleed brake.
Brake pull-off sluggish	Brake cam binding in housing	Free and grease.
	Weak brake shoe springs	Renew if springs have not become displaced.
	Sticking pistons in brake caliper (front disc brake only)	Overhaul caliper unit.
Harsh transmission	Worn or badly adjusted final drive chain	Adjust or renew as necessary.
	Hooked or badly worn sprockets	Renew as a pair.
	Loose rear sprocket (standard wheel only)	Check sprocket retaining bolts.

Chapter 8 Electrical system

Contents

Specifications

Battery (lead acid, 12 volt *)
Make	Lucas
Type	PUZ5A *

* Some models have two Lucas MK9E 6 volt batteries wired in series

Alternator
Make	Lucas
Type	RM19
	RM19ET (TR6C and T120TT)

Contact breaker ... ...   Lucas 4CA or 6CA

Rectifier type ... ...   Lucas 2DS 506

Zener diode type ... ...   Lucas ZD 715

Coils ... ...   Lucas MA12 - 2 off
Lucas 3ET - 2 off (TR6C and T120TT)

Ignition capacitor ... ...   Lucas 2MC

Fuse rating ... ...   35 amp

Bulbs
Headlamp	45/35W Pre-focus 12 volt Lucas 370
Pilot	6W, 12 volt Lucas 989
Stop and tail lamp	5/21W offset pin, 12 volt Lucas 380
Speedometer lamp	3W, 12 volt Lucas 987
Tachometer lamp	3W, 12 volt Lucas 987
Ignition warning lamp	2W, 12 volt Lucas 281
Main beam warning lamp	2W, 12 volt Lucas 281
Indicator warning lamp...	2W, 12 volt Lucas 281
Flashing indicator lamps	21W, 12 volt Lucas 382

1 General description -

An alternating current generator driven from the end of the crankshaft powers the electrical system. The output is converted into direct current by a silicon diode rectifier and supplied to a 12 volt battery. An electrical device, known as a Zener diode regulates the charge rate to suit the condition of the battery.

The ignition system, as described in Chapter 5, derives its supply from the rectified current. No emergency start facility is provided because even with a 'flat' battery, the generator output is sufficient to provide the necessary spark.

Some of the earlier coil ignition models (prior to engine number DU 24875) operate on 6 volts and have a different type of ignition switch which provides an emergency start facility. These models have a generator which will provide alternative charge rates, governed by the position of the lighting switch. A few machines, made during the transition stage from 6 to 12 volt operation, have two 6 volt batteries, connected in series.

2 Alternator - checking output

1   The output and performance of the alternator fitted to the Triumph 650/750 cc unit-construction twins can be checked only with specialised test equipment of the multi-meter type. It is unlikely that the average owner will have access to this type of equipment or instruction in its use. In consequence, if the performance is suspect, the alternator and charging circuit should be checked by a qualified auto-electrical expert.
2   Failure of the alternator does not necessarily mean that a replacement is needed. This can however sometimes be most economic through a service exchange scheme. It is possible to replace or rewind the stator coil assembly, for example, if the rotor is undamaged.
3   If the generator fails to charge, a warning light in the headlamp shell will indicate.

3 Battery - charging procedure and maintenance

1   Whilst the machine is used on the road it is unlikely that the battery will require attention other than routine maintenance because the generator will keep it fully charged. However, if the machine is used for a succession of short journeys only, mainly during the hours of darkness when the lights are in full use, it is possible that the output from the generator may fail to keep pace with the heavy electrical demand, especially if the machine is parked with the lights switched on. Under these circumstances, it will be necessary to remove the battery from time to time to have it charged independently.
2   The battery is located below the dual seat, in a carrier slung between the two parallel frame tubes. It is secured by a strap which, when released, will permit the battery to be withdrawn after disconnection of the leads. The battery positive is always earthed. To remove the battery carrier, release the earth lead and rectifier by unscrewing the retaining nut. Slacken the nuts on the cross-straps and lift the carrier away.
3   The normal charge rate is 1 amp. A more rapid charge can be given in an emergency, but this should be avoided if possible because it will shorten the life of the battery.
4   When the battery is removed from the machine, remove the cover and clean the battery top. If the terminals are corroded, scrape them clean and cover them with vaseline (not grease) to protect them from further attack. If a vent tube is fitted, make sure it is not obstructed and that it is arranged so that it will not discharge over any parts of the machine.
5   If the machine is laid up for any period of time, the battery should be removed and given a 'refresher' charge every six weeks or so, in order to maintain it in good condition.
6   When two 6 volt batteries are fitted, they must be connected in series with one another. The negative of one battery must go to the wiring harness and the positive of the OTHER to the frame or earth connection. The intermediate connection is made by joining the free negative terminal of one battery to the free positive terminal of the other.

4 Silicon diode rectifier - general

1   The silicon diode rectifier is bolted to a bracket attached to the rear of the battery carrier, beneath the dual seat. Its function is to convert the alternating current from the alternator to direct current which can be used to charge the battery and operate the ignition circuit.
2   The rectifier is deliberately placed in this location so that it is not exposed directly to water or oil and yet has free circulation of air to permit cooling. It should be kept clean and dry; the nuts connecting the rectifier plates should not be disturbed under any circumstances.
3   It is not possible to check whether the rectifier is functioning correctly without the appropriate test equipment. If performance is suspect, a Triumph agent or an auto-electrical expert should be consulted. Note that the rectifier will be destroyed if it is subjected to a reverse flow of current.
4   When tightening the rectifier securing nut, hold the nut at the other end with a spanner. Apart from the fact that the securing stud is sheared very easily if overtightened, there is risk of the plates twisting and severing their internal connections.

5 Zener diode - general

1   The Zener diode is found only in the 12 volt systems, where it is used to regulate the amount of rectified current supplied to charge the battery.
2   In use, the Zener diode becomes quite hot and it is customary to fit the diode to a 'heat sink' which will disperse this heat by means of radiating fins. On early models (prior to engine number DU 66246) the diode was mounted behind a protective shield, on the left hand side of the machine, immediately below the nose of the dual seat. From engine number DU 66246 onwards, the position of the Zener diode was transferred to the bottom yoke of the forks, immediately below the headlamp. This gave the benefit of improved airflow and better cooling.
3   Specialised test equipment is required for testing the Zener diode. If the diode appears to malfunction, a Triumph agent or an auto-electrical expert should be consulted. The Zener diode is removed by withdrawing the rubber plug in the forward end of the heat sink, which will expose the mounting nut. Remove the spade end connector to the diode first, then unscrew the mounting nut and withdraw the diode, leaving the heat sink in position.
4   Do not make any connection between the diode mounting and the heat sink, otherwise the heat transfer will be affected, resulting in failure of the diode. When replacing the diode, note that the mounting stud is of copper which will shear easily if over-stressed.

6 Headlamp - replacing bulbs and adjusting beam height

1   Two types of headlamp are fitted, depending on the year of manufacture of the machine. The pre 1971 models have a headlamp of conventional shape that utilises a reflector unit of the pre-focus type. Post 1970 machines are fitted with a 'short' headlamp, having a recessed back which carries the oil pressure warning, ignition warning and main beam indicator lamps. The reflector unit is of the pre-focus type.
2   To replace either the main bulb or the pilot lamp bulb, it is necessary to remove the front of the headlamp in order to gain access to the rear of the reflector unit. Slacken the screw at the top of the headlamp rim and pull the rim complete with reflector unit from the headlamp shell. To remove the contacts from the main bulb holder, press and twist the cover. The main bulb can then be extracted complete with its locating plate and the replacement inserted. The locating plate obviates the need for refocussing, when the bulb is changed. The bulb is locked in position by the contact cover, having a triple bayonet connection offset to ensure that the correct connections are made.
3   The pilot bulb holder is a push fit in the reflector shell. When the holder is withdrawn, the bulb can be released from its bayonet fitting by pressing downward and turning.
4   The main bulb is rated at 45/35W, 12 volts and the pilot lamp bulb at 6W, 12 volts.
5   After the reflector and headlamp rim have been replaced, beam height can be adjusted by slackening the two bolts which secure the headlamp shell to the forks or, if flashing indicators are fitted, by slackening the locknut around the arms which

3.2 Battery is located below dualseat, across frame

3.4 Vent pipe is fitted to minimise acid spillage

5.2 Zener diode is mounted on lower fork yoke

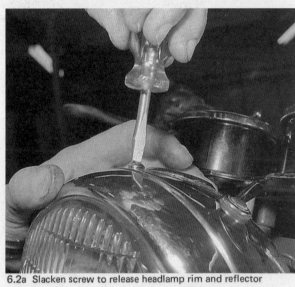

6.2a Slacken screw to release headlamp rim and reflector

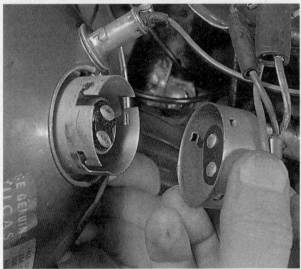

6.2b Press and twist cover to release bulb

6.2c Locating plate around bulb ensures correct focus

134

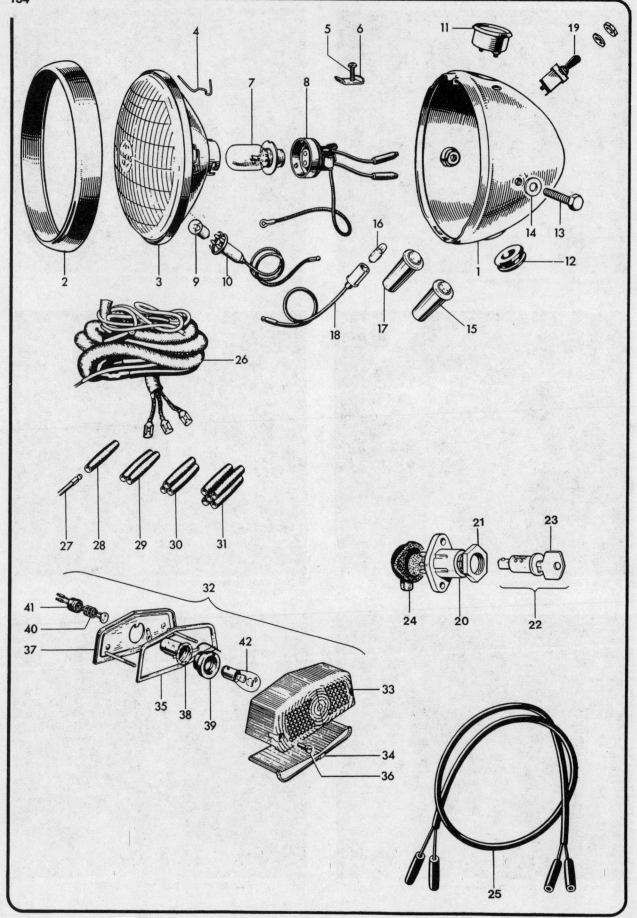

FIG. 8.1. HEADLAMP, WIRING HARNESS, SWITCHES AND TAIL LAMP

1	Headlamp complete, Lucas type SS700P	22	Ignition lock, complete with keys
2	Headlamp rim	23	Ignition key
3	Reflector unit	24	Ignition switch cover
4	Fixing wire clips - 6 off	25	Stop and tail lamp
5	Screw	26	Wiring harness
6	Plate	27	Snap connector terminal
7	Bulb (50/40W, vertical dip Pre-focus)	28	Single snap connector
8	Main bulbholder contacts	29	Double snap connector
9	Pilot bulb (6W. M.C.C.)	30	Triple snap connector
10	Pilot bulbholder	31	Quintuple snap connector
11	Ammeter	32	Stop and tail lamp, Lucas 564
12	Grommet	33	Stop lamp lens
13	Bolt - 2 off	34	Window
14	Plain washer - 2 off	35	Lens sealing gasket
15	Warning light body (red)	36	Sleeve nut - 2 off
16	Bulb (2W Lucas BA7S)	37	Base assembly
17	Warning light body (green)	38	Bulbholder
18	Warning light bulbholder - 2 off	39	Grommet
19	Lighting switch, Lucas 57SA	40	Contact assembly
20	Ignition switch, Lucas S45	41	Grommet
21	Nut	42	Bulb (6/21W, offset pin)

support the indicator lamps. Adjustments can now be made by tilting the headlamp upward or downward whilst the rider is seated normally. If a pillion passenger is carried, the adjustment made should take this into account because the headlamp will be raised as a result of the increased loading on the rear of the machine.

6 UK Lighting Regulations stipulate that the lighting system should be arranged so that the main beam will not dazzle a person standing in the same horizontal plane as the vehicle at a distance greater than 25 yards from the lamp, whose eye level is not less than 3 feet 6 inches above that plane. It is easy to approximate this setting by placing the machine 25 yards away from a wall, on a level surface, and setting the beam height so that the beam is concentrated at the same height as the distance from the centre of the headlamp to the ground. The rider must be seated normally during this operation and the pillion passenger, if one is carried regularly.

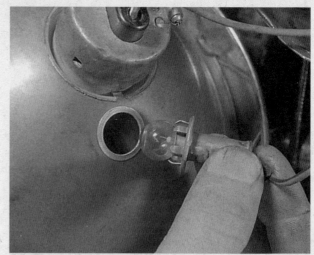

6.3 Pilot bulb holder is push fit in reflector

7 Tail and stop lamp - replacing bulb

1 The combined tail and stop lamp is fitted with a double filament bulb with offset pins to prevent its unintentional reversal in the bulb holder. The lamp unit serves a two-fold purpose; to illuminate the rear of the machine and the rear number plate, and to give visual warning when the rear brake is applied. To gain access to the bulb, remove the two screws securing the plastic lens cover in position. This has a sealing gasket below it. The bulb is released by pressing inward with a twisting action; it is rated at 5/21W, 12 volts.

2 The stop lamp is actuated by a switch bolted to the rear chainguard. The switch is connected to the brake rod by a spring attached to a clamp fitting around the rod; the point at which the switch operates is governed by the position of the clamp in relation to the rod. If the clamp is moved forward, the stop lamp will indicate earlier and vice versa. The switch does not require attention other than the occasional drop of thin oil.

7.1a Remove plastics lens cover for access to bulb

8 Speedometer and tachometer bulbs - replacement

The speedometer and tachometer heads are each fitted with a bulb to illuminate the dial when the headlamp is switched on. The bulb holders are a push fit into the bottom of each instrument case and carry a 3W, 12 volt bulb which has a threaded body.

9 Ignition warning, oil pressure warning, main beam and flashing indicator warning bulbs - replacement

1 The combination of lamps fitted varies according to the model and year of manufacture. There is not necessarily four separate forms of warning fitted to each machine.

2 The bulb holders are a push fit into either the top or the back of the headlamp shell, depending on the type of headlamp fitted. Each of the bulbs is rated at 2W, 12 volts.

10 Flashing indicator lamps

1 Late models have flashing direction indicator lamps attached to the front and rear of the machine. They are operated by a thumb switch on the left hand end of the handlebars. An indicator lamp built into the rear of the headlamp shell will flash in unison with the lights, provided the front and rear lights are operating correctly.

2 The bulbs are fitted by removing the plastic lens covers, held in position by two screws. The bulbs are of the bayonet type and must be pressed and turned to release or fit. Each bulb is rated at 21W, 12 volts.

7.1b Bulb has offset pins to prevent reversal of stop and tail functions

11 Flasher unit - location and replacement

1 The flasher unit is located beneath the dual seat, along with the other electrical equipment. It seldom gives trouble unless it is subjected to a heavy blow which will disturb its senstive action.
2 It is not possible to renovate a malfunctioning flasher unit. If the bulbs are in working order and will give only a single flash when the handlebar switch is operated, the flasher unit should be suspected and, if necessary, renewed.

12 Headlamp dip switch

1 The headlamp dip switch forms part of the switch unit fitted to the right hand side of the handlebars on all late models. Earlier models have a separate dip switch mounted on the left hand side of the handlebars which also contains the horn push.
2 If the dip switch malfunctions, the switch unit must be renewed since it is seldom practicable to effect a satisfactory repair.

13 Horn push and horn - adjustment

1 The horn push on late models forms part of the switch unit at the right hand end of the handlebars. On earlier models, it is combined with the separate dip switch.
2 The horn is secured below the nose of the petrol tank, facing in a forward direction. It is provided with adjustment in the form of a serrated screw inset into the back of the horn body.
3 To adjust the horn, turn the screw anticlockwise until the horn just fails to sound, then back it off about one-quarter turn. Adjustment is needed only very occasionally, to compensate for wear of the internal moving parts.

14 Fuse - location and replacement

1 A fuse is incorporated in the brown/blue coloured lead from the negative terminal of the battery. It is housed within a quickly detachable shell and protects the electrical equipment from accidental damage if a short circuit should occur.
2 If the electrical system will not operate, a blown fuse should be suspected, but before the fuse is renewed, the electrical system should be inspected to trace the reason for the failure of the fuse. If this precaution is not observed, the replacement fuse may blow too.
3 The fuse is rated at 35 amps and at least one spare should always be carried. In an extreme emergency, when the cause of the failure has been rectified and if no spare is available, a get-you-home repair can be effected by wrapping silver paper around the blown fuse and re-inserting it in the fuse holder. It must be stressed that this is only an emergency measure and the 'bastard' fuse should be replaced at the earliest possible opportunity. It affords no protection whatsoever to the electrical circuit when bridged in this fashion.

15 Ignition switch

1 The ignition switch is fitted to the left hand top cover of the forks, or in the left hand or right hand cover surrounding the rear portion of the frame, depending on the model and year of manufacture.
2 It is retained by a locknut or a locking ring which, when unscrewed, will free the switch.

16 Headlamp switch

1 A two or three position headlamp switch is fitted to the headlamp shell to operate the pilot and main headlamp bulbs. Machines fitted with the two-position rotary switch must have the ignition switch in position 4 before the headlamp will operate.
2 Late models have an additional headlamp flasher in the form of a push button embodied in the switch assembly on the right hand end of the handlebars.

17 Ammeter

1 All models with the full headlamp shell (except those fitted with AC ignition) have an ammeter inset into the top of the headlamp shell to show the amount of charge from the generator.
2 When an ammeter is not fitted, as in the case of the late models with the 'short' headlamp shell, an ignition warning light is fitted. This light will not extinguish after the engine is started, if the generator has failed.

18 Capacitor ignition

1 A capacitor is built into the ignition circuit so that a machine without lights (and therefore without a battery) or one on which the battery has failed, can be started and run normally. If lights are fitted, the machine can be used during the hours of darkness, since the lighting equipment will function correctly immediately the engine starts, even if the battery is removed.
2 The capacitor is fitted into a coil spring to protect it from vibration. It is normally mounted with its terminals pointing downward from a convenient point underneath the dual seat.
3 Before running a machine on the capacitor system with the battery disconnected, it is necessary to tape up the battery negative so that it cannot reconnect accidentally and short circuit. If this occurs, the capacitor will be ruined. A convenient means of isolating the battery is to remove the fuse.

19 Wiring - layout and examination

1 The cables of the wiring harness are colour-coded and will correspond with the accompanying wiring diagrams.
2 Visual inspection will show whether any breaks or frayed outer coverings are giving rise to short circuits which will cause the main fuse to blow. Another source of trouble is the snap connectors and spade terminals, which may make a poor connection if they are not pushed home fully.
3 Intermittent short circuits can sometimes be traced to a chafed wire passing through, or close to, a metal component, such as a frame member. Avoid tight bends in the cables or situations where the cables can be trapped or stretched, especially in the vicinity of the handlebars or steering head.

20 Front brake stop lamp switch

1 In order to comply with traffic requirements in certain overseas countries, a stop lamp switch is now incorporated in the front brake cable so that the rear stop lamp is illuminated when the front brake is applied. When an hydraulic disc brake is fitted, the stop lamp switch is incorporated in the master cylinder unit, clamped to the right hand end of the handlebars.
2 There is no means of adjustment for the front brake stop lamp switch. If the switch malfunctions, the front brake cable must be replaced, or in the case of the disc brake assembly, the switch unit unscrewed from the master cylinder and renewed.

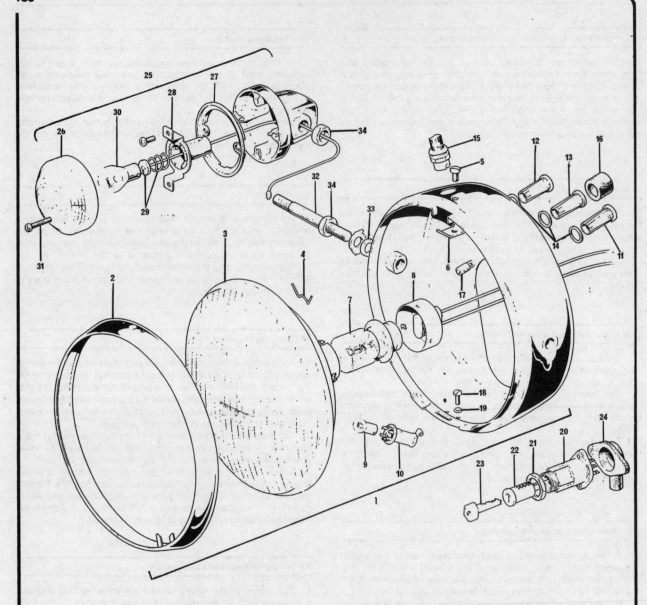

FIG. 8.2. POST 1970 - HEADLAMP AND FLASHING INDICATORS

1 Headlamp complete, Lucas MCH 69	11 Warning light (red)	20 Ignition switch, Lucas 149SA	29 Interior assembly - 4 off
2 Headlamp rim	12 Warning light (green)	21 Nut	30 Bulb - 4 off
3 Reflector unit	13 Warning light (amber)	22 Ignition lock, complete with key	31 Screw - 8 off
4 Fixing wire clips - 6 off	14 Sealing washer - 3 off	23 Key	32 Stanchion, flashing indicator lamps - 4 off
5 Screw	15 Light switch	24 Cover, ignition switch	33 Washer - 2 off
6 Plate	16 Shield, warning light	25 Flashing indicator lamp - 4 off	34 Locknut - 8 off
7 Main bulb (45/35W)	17 Bulb, warning light - 3 off	26 Lens - 4 off	
8 Main bulbholder contacts	18 Screw - 2 off	27 Gasket - 4 off	
9 Pilot bulb	19 Washer - 2 off	28 Bulbholder - 4 off	
10 Pilot bulbholder			

WIRING DIAGRAMS

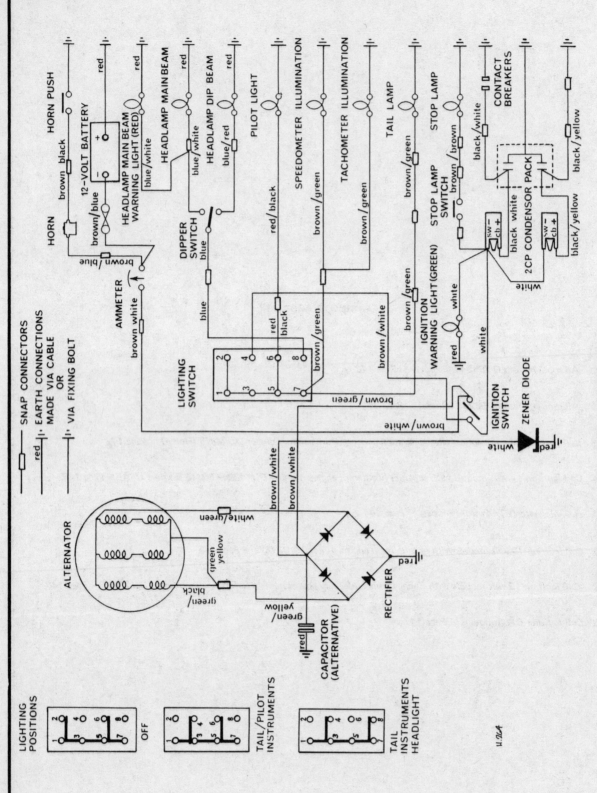

Wiring Diagram 1. All models from DU66246 (Export)

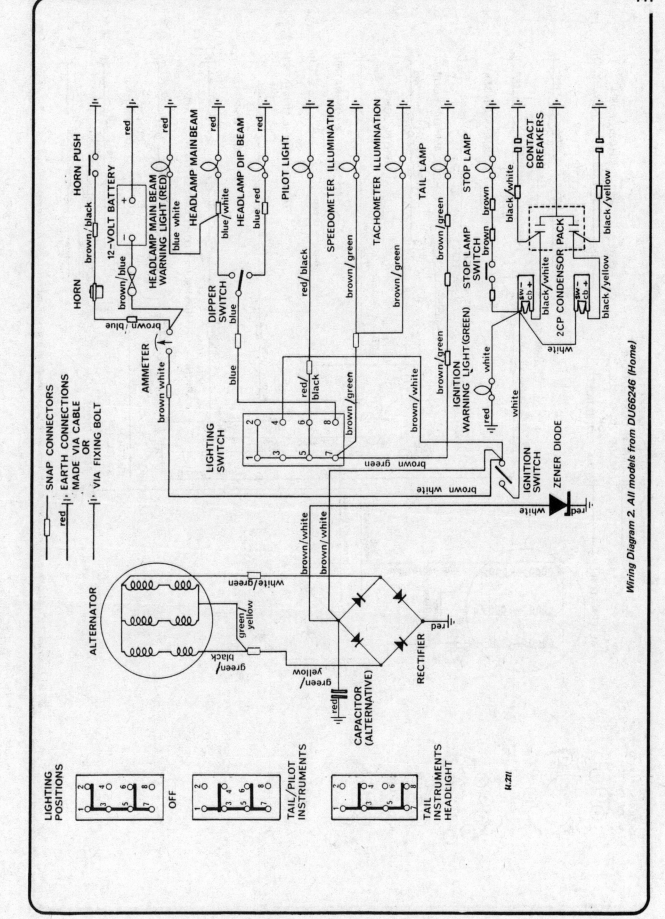

Wiring Diagram 2. All models from DU66246 (Home)

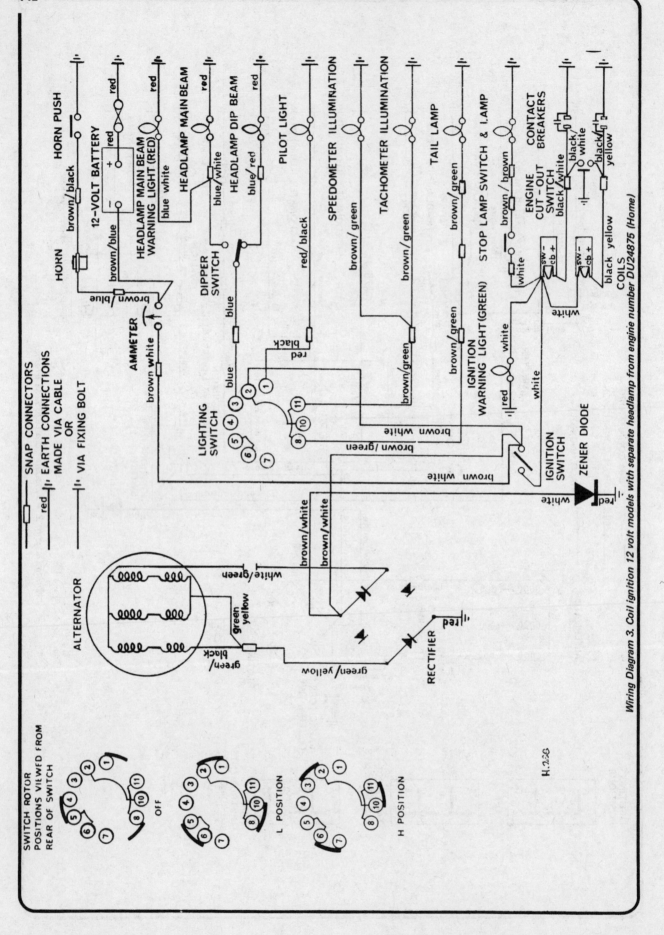

Wiring Diagram 3. Coil ignition 12 volt models with separate headlamp from engine number DU24875 (Home)

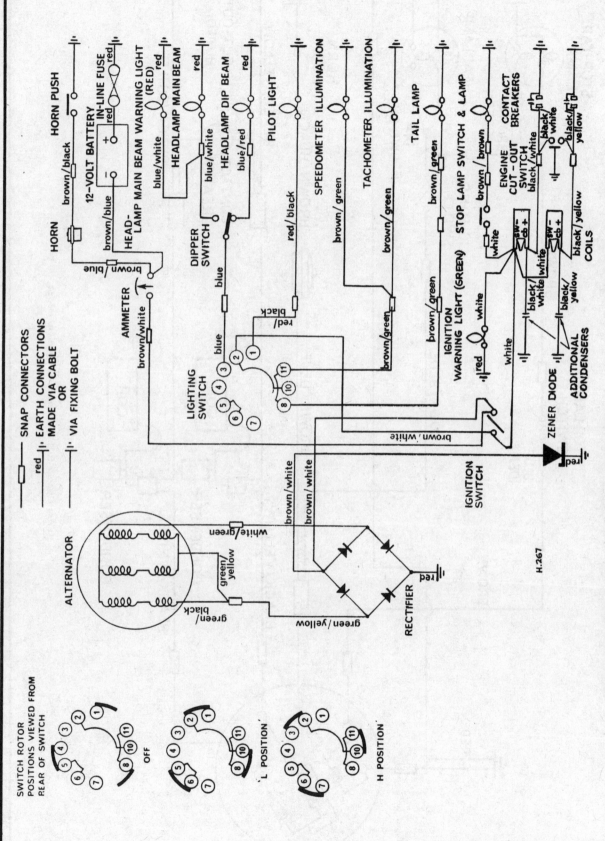

Wiring Diagram 4. Coil ignition 12 volt models with separate headlamp, engine number DU44394–66245 (Export USA)

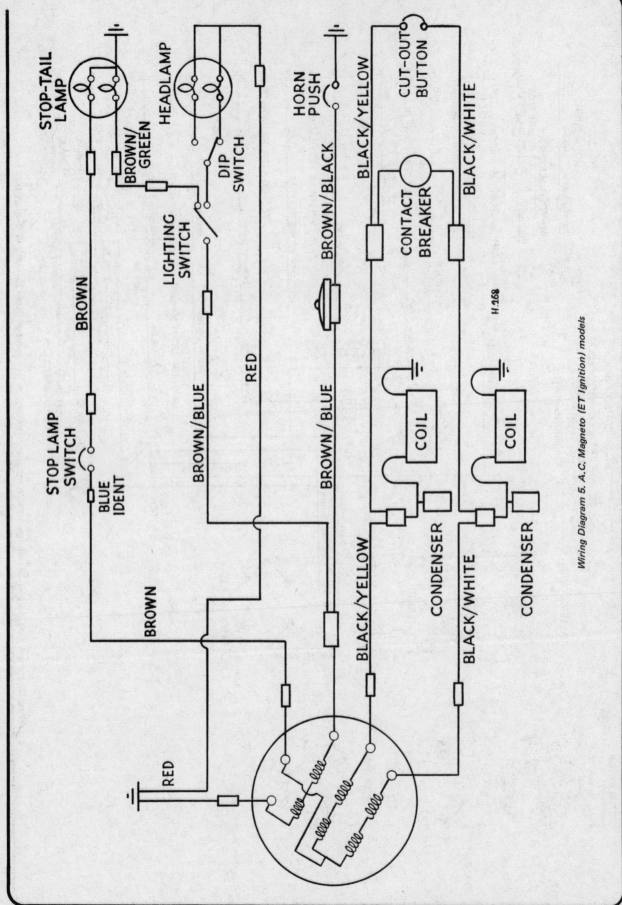

Wiring Diagram 5. A.C. Magneto (ET Ignition) models

STOP-TAIL LAMP

HEADLAMP

HORN PUSH

CUT-OUT BUTTON

BROWN/GREEN

DIP SWITCH

BROWN/BLACK

BLACK/YELLOW

CONTACT BREAKER

BLACK/WHITE

BROWN

LIGHTING SWITCH

RED

BROWN/BLUE

COIL

COIL

STOP LAMP SWITCH

BLUE IDENT

BROWN/BLUE

CONDENSER

CONDENSER

H.268

BROWN

RED

BLACK/YELLOW

BLACK/WHITE

145

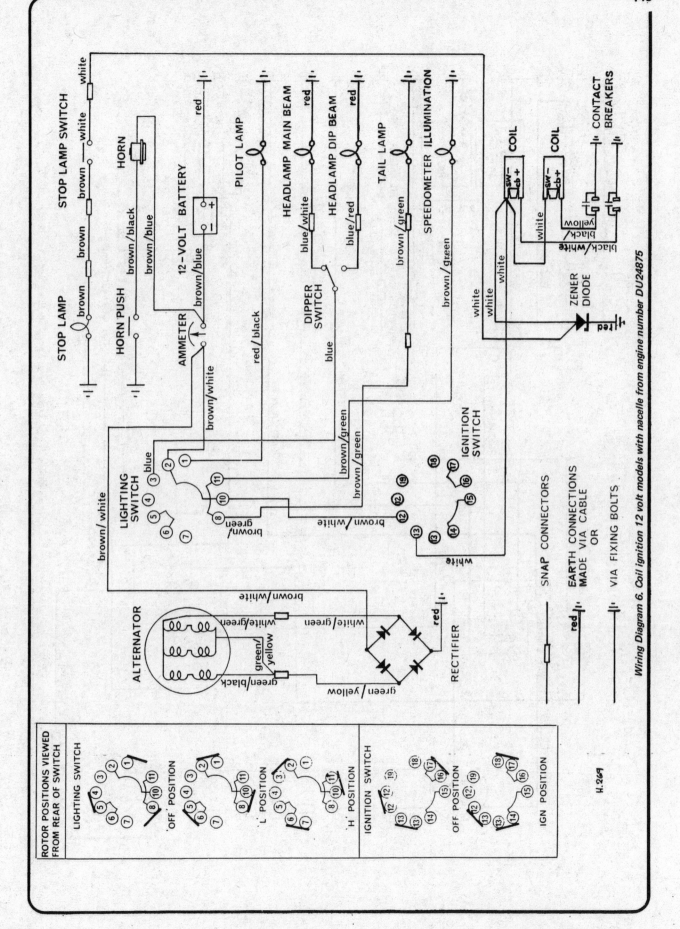

Wiring Diagram 6. Coil ignition 12 volt models with nacelle from engine number DU24875

H.269

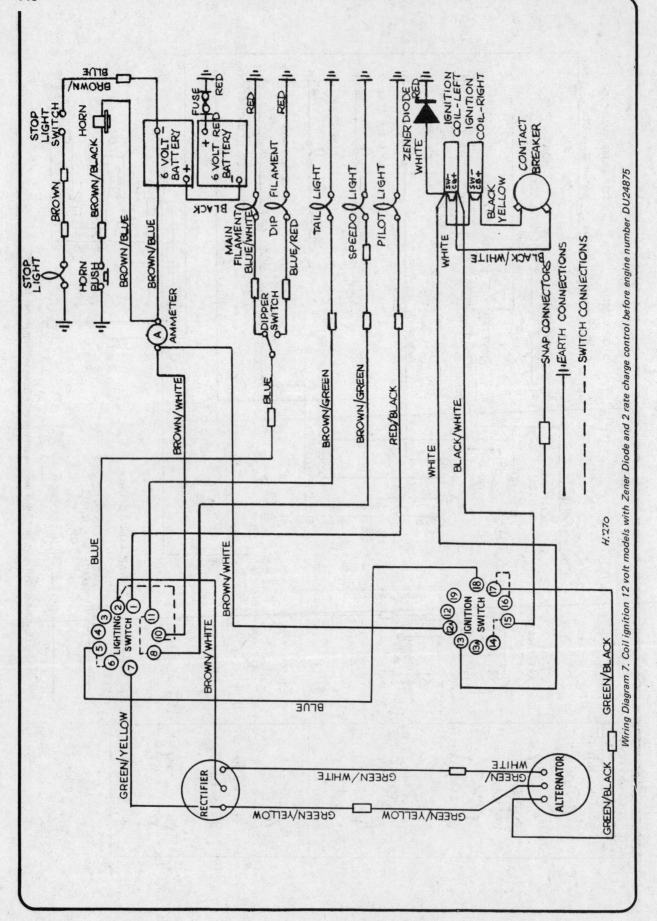

Wiring Diagram 7. Coil ignition 12 volt models with Zener Diode and 2 rate charge control before engine number DU24875

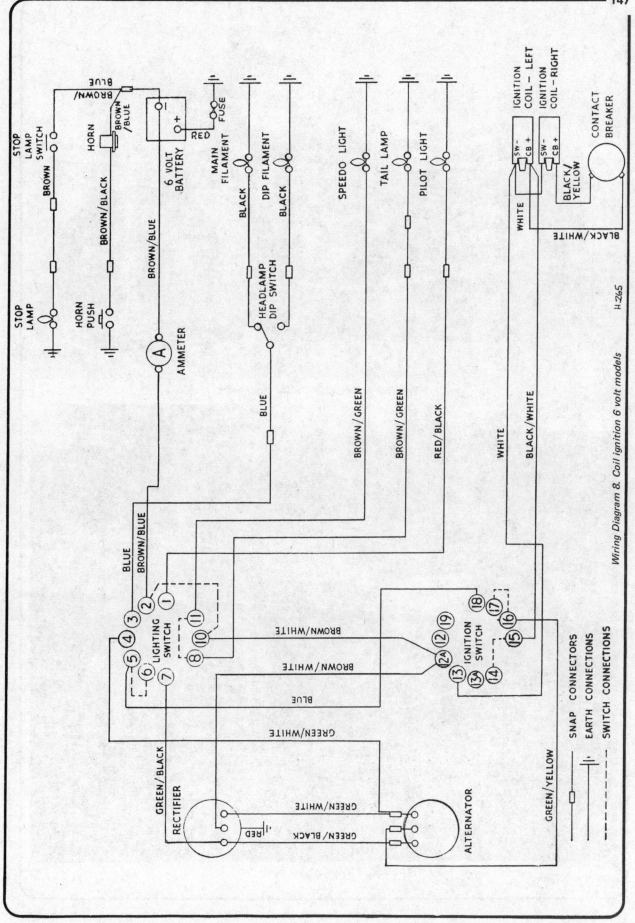

Wiring Diagram 8. Coil ignition 6 volt models

The 750 cc Triumph T140V Bonneville

The 750 cc Triumph Jubilee

The 750 cc Triumph T140 D Special

The 650 cc Triumph TR65 Thunderbird

The 750 cc Triumph T140 ES Bonneville Electro

The 750 cc Triumph TR7T Tiger Trail

Chapter 9 The 1975 to 1983 models

Contents

Specifications

Note: At the time of publication comprehensive specificational information for the TR65T model was not readily available. Every effort, however, has been made to include it where possible.

Specifications relating to Chapter 1

Engine		650 models	750 models
Bore and stroke		76 x 71 mm ... ... (2.993 x 2.795 in) ...	76 x 82 mm (2.993 x 3.228 in)
Cubic capacity		649 cc ... ... (39.6 cu in)	744 cc (45 cu in)
Compression ratio			
TR65		8.6 :1	N/App
TR65T		7.5 : 1	N/App
TR7T		N/App	7.4 : 1
All other models		N/App	7.9 : 1

152

Crankshaft

Type ... ...   Forged 2-throw crank. Bolt-on flywheel.
Located by timing side main bearings.

Main bearing size and type:
  Drive side ... ...   Single lipped roller -
71.5 x 28.6 x 20.6 mm
(2.812 x 1.125 x 0.812 in)

  Timing side:
    Pre EDA 30 000 ... ...   Single row ball — 72 x 30 x 19 mm
(2.834 x 1.181 x 0.748 in)

    Post EDA 30000 ....   Twin lipped roller — 72 x 30 x 19 mm
(2.834 x 1.181 x 0.748 in)

Main bearing journal diameter:
  Drive side ... ...   28.576 x 28.575 mm
(1.1247 - 1.1250 in)

  Timing side ... ...   30.0 - 29.99 mm
(1.1812 - 1.1808 in)

Big end journal diameter ... ...   41.237 - 41.250 mm
(1.6235 - 1.6240 in)

Minimum regrind diameter ... ...   40.730 - 40.742 mm
(1.6035 - 1.6040 in)

Crankshaft endfloat ... ...   0.076 - 0.432 mm
(0.003 - 0.017 in)

Oil feed diameter ... ...   15.798 - 15.810 mm
(0.622 - 0.623 in)

Connecting rods

Length between centres... ... ...   152.425 - 152.375 mm
(6.001 - 5.999 in)

Big end diameter... ... ...   273.568 - 273.553 mm
(1.770 - 1.769 in)

Small end diameter ... ...   19.085 - 19.077 mm
(0.751 - 0.750 in)

Big end bearings

Type ... ...   White metal, steel backed shells
Side clearance ... ...   0.305 - 0.406 mm
(0.012 - 0.016 in)

Radial clearance ... ...   0.013 - 0.051 mm
(0.0005 - 0.0020 in)

Cylinder head

Material type ... ...   Aluminium alloy
Valve seat angle ... ...   45°
Valve seat interference ... ...   0.1016 mm (0.004 in)
Valve guide bore ... ...   12.66 - 12.64 mm

Cylinder block

Material type ... ...   Cast iron
Standard bore:
  Low (L) grade ... ...   75.992 - 75.979 mm
(2.9918 - 2.9913 in)

  High (H) grade ... ...   76.007 - 75.994 mm
(2.9924 - 2.9919 in)

Maximum oversize bore... ... ...   76.979 - 76.951 mm
(3.0321 - 3.0310 in)

Tappet guide block housing
diameter ... ...   25.370 - 25.362 mm
(0.9990 - 0.9985 in)

Tappet guide block

Outer diameter ... ...   25.40 - 25.387 mm (1.0 - 0.9995 in)
Bore(s) diameter ... ...   7.925 - 7.938 mm (0.3120 - 0.3125 in)
Interference fit in cylinder block ... ...   0.0127 - 0.267 mm (0.0005 - 0.0015 in)

Tappets

Type ... ...   Stellite tip. High tensile steel body
Diameter ... ...   7.90 - 7.913 mm (0.3110 - 0.3115 in)
Clearance in guide block ... ...   0.0127 - 0.0381 mm (0.0005 - 0.0015 in)
Tip radius:
  Inlet ... ...   19.10 mm (0.75 in)

Exhaust:
    Up to and including
    1978 models ... ... 28.56 mm (1.125 in)
    Post 1978 models ... ... 19.10 mm (0.75 in)

Valves
Head diameter:
    Inlet ... ... 40.437 - 40.538 mm (1.592 - 1.596 in)
    Exhaust ... ... 36.388 - 36.576 mm (1.434 - 1.440 in)
Stem diameter:
    Inlet ... ... 7.860 - 7.874 mm (0.3095 - 0.3100 in)
    Exhaust ... ... 7.849 - 7.861 mm (0.3090 - 0.3905 in)

Valve springs
Inner (Red spot):
    Total no of coils 7.25
    Free length (new) .... 38.90 mm (1.17/32 in)
Outer (Green spot):
    Total no of coils 5.50
    Free length (new) ... ... 41.30 mm (1.5/8 in)

Valve guides
Type ... ... Aluminium/bronze
Outer diameter (inlet and exhaust) ... ... 12.713 - 12.725 mm (0.5005 - 0.5010 in)
Bore diameter (inlet and exhaust):
    Up to and including
    1978 models ... ... 7.943 - 7.958 mm (0.3127 - 0.3137 in)
    Post 1978 models ... ... 7.925 - 7.950 mm (0.3120 - 0.3130 in)
Length:
    Inlet ... ... 50 mm (1 31/32 in)
    Exhaust ... ... 55.2 mm (2 11/64 in)

Valve timing
With nil tappet clearance:
Valve lift:
    Inlet opens ... ... 3.83 mm (0.151 in) 4.85 mm (0.190 in)
        at TDC   at TDC
    Exhaust closes ... ... 2.23 mm (0.088 in) 3.27 mm (0.130 in)
        at TDC   at TDC

Pushrods
Overall length ... ... 143.40 - 144.11 mm 149.27 - 149.98 mm
        (5.646 - 5.674 in)   (5.877 - 5.905 in)

Rockers
Type ... ... High tensile steel forging
Spindle diameter... ... ... 12.675 - 12.687 mm (0.499 - 0.4995 in)
Bore diameter ... ... 12.705 - 12.731 mm (0.5002 - 0.5012 in)
Tappet clearance (cold):
    Inlet ... ... 0.203 mm (0.008 in)
    Exhaust ... ... 0.15 mm (0.006 in)

Pistons
Type ... ... Diecast - aluminium alloy
Diameter:
    Low (L) grade ... ... 75.872 - 75.885 mm (2.9871 - 2.9876 in)
    High (H) grade ... ... 75.887 - 75.900 mm (2.9877 - 2.9882 in)
Gudgeon pin hole diameter ... ... 19.051 - 19.060 mm (0.7502 - 0.7504 in)
Gudgeon pin diameter ... ... 19.050 - 19.055 mm (0.7500 - 0.7502 in)
Oversizes available ... ... + 0.254 mm (0.010 in)
        + 0.508 mm (0.020 in)
        + 0.762 mm (0.030 in)
        + 1.016 mm (0.040 in)

Piston rings
Compression rings (tapered):
    Thickness ... ... 1.589 - 1.563 mm (0.0625 - 0.0165 in)
    Width ... ... 3.073 - 2.896 mm (0.121 - 0.113 in)
    Clearance in groove ... ... 0.089 - 0.038 mm (0.0035 - 0.0015 in)
    Fitted gap ... ... 0.203 - 0.330 mm (0.008 - 0.013 in)

Oil control ring:
Thickness ... ... 3.18 mm (0.125 in)
Width ... ... 3.073 mm (0.121 in)
Clearance in groove ... ... 0.038 - 0.0635 mm (0.0015 - 0.0025 in)
Fitted gap:
Up to and including
1978 models ... ... 0.254 - 1.016 mm   (0.010 - 0.040 in)
Post 1978 models ... ... 0.203 - 0.254 mm (0.008 - 0.010 in)

Camshafts
Endfloat ... ... 0.331 - 0.508 mm   (0.013 - 0.020 in)
Journal diameter:
Left ... ... 20.564 - 20.577 mm (0.8100 - 0.8105 in)
Right ... ... 22.174 - 22.187 mm (0.8730 - 0.8735 in)
Diametral clearance:
Left ... ... 0.0254 - 0.0635 mm (0.0010 - 0.0025 in)
Right ... ... 0.0127 - 0.0508 mm (0.0005 - 0.0020 in)

Camshaft bearing bushes
Type ... ... High density sintered bronze
Outer diameter:
Left ... ... 25.425 - 25.438 mm (1.0010 - 1.0015 in)
Right ... ... 28.601 - 28.628 mm (1.126 - 1.127 in)
Bore diameter (fitted):
Left ... ... 20.648 - 20.663 mm (0.8125 - 0.8135 in)
Right ... ... 22.199 - 22.225 mm (0.874 - 0.875 in)
Length
Inlet (left) ... ... 28.041 - 28.296 mm (1.104 - 1.114 in)
Exhaust (left) ... ... 23.637 - 23.927 mm (0.932 - 0.942 in)
Inlet (right) ... ... 25.025 - 25.508 mm (1.010 - 1.020 in)
Exhaust (right) ... ... 25.025 - 25.508 mm (1.010 - 1.020 in)
Interference fit in crankcase:
Left ... ... 0.025 - 0.051 mm (0.001 - 0.002 in)
Right ... ... 0.025 - 0.064 mm (0.001 - 0.0025 in)

Timing gears (kickstart models)
Crankshaft pinion:
No of teeth 25
Fit on crankshaft ... ... + 0.0076/ - 0.0127 mm (0.0003/0.0005 in)
Intermediate timing gear:
No of teeth 47
Bore diameter ... ... 14.27 - 14.188 mm (0.5618 - 0.5625 in)
Intermediate timing gear bush:
Type ... ... Phosphor bronze
Outer diamter ... ... 14.313 - 14.326 mm (0.5635 - 0.5640 in)
Bore diameter ... ... 12.675 - 12.687mm (0.4990 - 0.4995 in)
Working clearance on spindle ... ... 0.0127 - 0.267 mm (0.0005 - 0.0015 in)
Length ... ... 17.209 - 17.336 mm (0.6775 - 0.6825 in)
Inlet and exhaust camshaft pinions:
No of teeth 50
Interference fit on camshaft ... ... 0.000 - 0.0254 mm (0.000 - 0.001 in)
Intermediate wheel spindle :
Diameter ... ... 12.649 - 12.662 mm (0.4980 - 0.4985 in)
Interference fit in crankcase ... ... 0.0127 - 0.267 mm (0.0005 - 0.0015 in)

Timing gears (electric start models)
Crankshaft pinion:
No of teeth 25
Fit on crankshaft ... ... + 0.0076/ - 0.0127 mm (0.0003/0.0005 in)
Intermediate timing gear:
No of teeth 47
Bore diameter:
At clutch ... ... 48.399 - 48.425 mm (1.9055 - 1.9065 in)
At spindle ... ... 14.270 - 14.288 mm (0.5618 - 0.5625 in)
Intermediate timing gear spindle
bush:
Pre engine no 33065:
Outer diameter ... ... 15.913 - 15.925 mm (0.6265 - 0.6270 in)
Bore diameter ... ... 12.70 - 12.713 mm (0.50 - 0.5005 in)
Length ... ... 17.208 - 17.335 mm (0.6775 - 0.6825 in)
Interference fit in crankcase ... ... 0.040 mm (0.0015 in)

Post engine no 33065:

Outer diameter	21.038 - 21.051 mm (0.8283 - 0.8288 in)
Bore diameter	12.70 - 12.713 mm (0.50 - 0.5005 in)
Length	17.208 - 17.335 mm (0.6775 - 0.6825 in)
Interference fit in crankcase	0.1016 mm (0.004 in)

Inlet and exhaust camshaft pinions:

No of teeth	50
Interference fit on camshaft	0.000 - 0.0254 mm (0.000 - 0.001 in)

Intermediate timing gear spindle

diameter	12.651 - 12.669 mm (0.4981 - 0.4988 in)

Starter drive (electric start models)

Clutch hub bush:

Outer diameter	31.745 - 31.758 mm (1.2498 - 1.2503 in)
Bore diameter	19.045 - 19.063 mm (0.7498 - 0.7505 in)

Clutch hub bush

Outer diameter	19.088 - 19.010 mm (0.7515 - 0.7520 in)
Bore diameter	12.708 - 12.720 mm (0.5003 - 0.5008 in)
Length	17.272 mm (0.680 in)

Starter gear nos 4 and 6:

Bore diameter	12.786 - 12.802 mm (0.5034 - 5040 in)
Shaft diameter	12.740 - 12.758 mm (0.5016 - 0.5023 in)

Torque wrench settings

	lbf ft	kgf m
Flywheel bolts	33	4.6
Connecting rod bolts	22	3.0
Crankcase junction bolts	13	1.8
Crankcase junction studs	20	2.8
Rocker box bolts:		
5/16 in diameter (inner)	10	1.38
¼ in diameter	5	0.7
Rocker box nuts	5	0.7
Rocker spindle domed nuts	22	3.0
Cylinder head bolts - 3/8 in diameter (outer)	18	2.49
Cylinder head bolts - 5/16 in diameter (centre)	16	2.2
Cylinder head bolts - 3/8 in diameter (inner)...	18	2.49
Primary cover domed nuts	10	1.4
Cylinder block base nuts	20	2.8
Crankshaft pinion nut	80	11.0
Crankshaft sprocket nut	50	7.0

Specifications relating to Chapter 2

Gearbox

Internal ratios:

5th	1.00 : 1
4th	1.19 : 1
3rd	1.40 : 1
2nd	1.837 : 1
1st	2.585 : 1

Overall ratios (Pre EDA 30 000):

5th	4.70 : 1
4th	5.59 : 1
3rd	6.58 : 1
2nd	8.63 : 1
1st	12.25 : 1 (12.14 : 1 for post 1978 models)
Gearbox sprocket teeth	20
Engine rpm at 10 mph in 5th gear	627 (669 for post 1978 models)

Overall ration (Post EDA 30 000)

	TR65	TR65T and TR7T	T140 and TR7
5th	4.95 : 1	5.22 : 1	4.50 : 4.50 : 1
4th	5.89 : 1	6.21 : 1	5.36 : 1
3rd	6.93 : 1	7.31 : 1	6.30 : 1
2nd	9.09 : 1	9.59 : 1	8.27 : 1
1st	12.79 : 1	13.49 : 1	11.63 : 1
Gearbox sprocket teeth	19	18	20
Engine rpm at 10 mph in 5th gear	660	680	640

No of pinion teeth (mainshaft/layshaft):

5th	21/15
4th	20/17
3rd	18/18
2nd	16/21
1st	13/24

Mainshaft high gear

Bearing:

Type	Needle roller (Torrington B1314)
Length	22.230 - 21.970 mm (0.875 - 0.865 in)

Mainshaft

Left end diameter	20.612 - 20.625 mm (0.8115 - 0.8120 in)
Right end diameter	19.044 - 19.054 mm (0.7494 - 0.7498 in)
Length	286.258 mm (11.270 in)

Layshaft

Left end diameter	17.460 - 17.404 mm (0.6875 - 0.6870 in)
Right end diameter	17.460 - 17.404 mm (0.6875 - 0.6870 in)
Length	164.33 mm (6.47 in)
End float:	
Minimum	0.0635 mm (0.0025 in)
Maximum	1.20651.2065 mm (0.0475 in)

Bearings

Mainshaft bearing:

Left	38.1 x 63.5 x 15.9 mm (1.5 x 2.5 x 0.625 in) Roller bearing
Right	19.0 x 47.5 x 14.3 mm (0.75 x 1.875 x 0.563 in) Ball journal

Layshaft bearing:

Left	17.5 x 22.23 x 19.0 mm (0.0688 x 0.875 x 0.75 in) Needle roller
Right	17.5 x 22.23 x 19.0 mm (0.0688 x 0.875 x 0.75 in) Needle roller

Layshaft 1st gear bush:

Bore diameter	20.320 - 20.203 mm (0.800 - 0.795 in)
Shaft diameter	20.511 - 20.498 mm (0.8075 - 0.8070 in)

Layshaft 2nd gear bush:

Bore diameter	20.320 - 20.203 mm (0.800 - 0.795 in)
Shaft diameter	20.511 - 20.498 mm (0.8075 - 0.8070 in)

Gearchange mechanism

Outer quadrant:

Bush bore diameter	15.860 - 15.888 mm (0.6245 - 0.6225 in)
Clearance on shaft	0.0178 - 0.081 mm (0.0007 - 0.0032 in)

Inner quadrant (pre 1978 models):

Bush bore diameter	19.063 - 19.075 mm (0.7505 - 0.7510 in)
Clearance on shaft	0.0127 - 0.064 mm (0.0005 - 0.0025 in)

Quadrant plungers:

Outer diameter	10.920 - 10.937 mm (0.4315 - 0.4320 in)
Working clearance in bore	0.0127 - 0.038 mm (0.0005 - 0.0015 in)

Quadrant plunger springs:

Free length	31.75 mm (1.25 in)
No of coils	12

Quadrant return springs:

Free length	44.50 mm (1.75 in)
No of coils:	
Up to and including 1978 models	9.50
Post 1978 models	12.0

Complete plunger:

Plunger length	43.656 mm (1.719 in)
Plunger diameter	11.06 - 11.09 mm (0.4355 - 0.4365 in)
Spring free length	54.356 mm (2.14 in)
No of coils	19

Gearchange shaft (post 1980 models only):

Shaft diameter:

Left-hand outer	18.97 - 19.03 mm (0.747 - 0.749 in)
Left-hand inner	18.95 - 18.97 mm (0.746 - 0.747 in)
Right-hand outer	15.794 - 15.844 mm (0.6218 - 0.6238 in)
Right-hand inner	18.97 - 19.03 mm (0.747 - 0.749 in)

Bush bore diameter (post 1980 models unless otherwise stated):

Left-hand outer	19.101 - 19.126 mm (0.752 - 0.753 in)
Left-hand inner	19.063 - 19.075 mm (0.7505 - 0.7510 in)

Right-hand outer	...	...	...	...	...	15.887 - 15.90 mm (0.6255 - 0.6260 in)
						— pre 1980 models after engine no HN62501,
						19.10 - 19.11 mm (0.7520 - 0.7525 in)
Right-hand inner	...	...	...	...	...	19.063 - 19.075 mm (0.7505 - 0.7510 in)
Gearbox oil capacity	...	...	...	...	...	500 cc (17.6/16.9 Imp/US fl oz)

Kickstart operating mechanism

Bush bore diameter	...	...	...	...	...	19.085 - 19.110 mm (0.751 - 0.752 in)
Spindle working clearance in bush	...	...	...	...	0.076 - 0.127 mm (0.003 - 0.005 in)
Ratchet spring free length	...	...	...	...	12.70 mm (0.50 in)

Torque wrench settings

						lbf ft	kgf m
Gearbox sprocket nut	...	...	...	...	...	80	11.0
Kickstart ratchet pinion nut	...	...	...	...	35	4.8

Specifications relating to Chapter 3

Clutch

Type	...	....	...	...	...	...	....	...	Multiplate with integral shock absorber
No of plates:									
Driven (plain):									
Pre 1980 models	...	...	...	...	...	...	6
Post 1980 models	...	...	...	...	...	7
Driving (bonded)	...	...	...	...	...	...	6
Clutch hub bearing diameter	...	...	...	...	33.882 - 34.907 mm (1.3733 - 1.3743 in)
Clutch sprocket bore diameter	...	...	...	...	47.612 - 47.638 mm (1.8745 - 1.8755 in)
Thrust washer thickness	...	...	...	...	...	1.312 - 1.372 mm (0.052 - 0.054 in)
Clutch sprocket teeth	...	...	...	...	...	58
Engine sprocket teeth	...	...	...	...	...	29
Bearing rollers:									
Number	...	...	...	...	...	...	...	20
Diameter	...	...	...	...	...	...	6.337 - 6.350 mm (0.2495 - 0.250 in)
Length	...	...	...	...	...	...	5.831 - 5.958 mm (0.231 - 0.236 in)
Pressure springs:									
Number	...	...	...	...	...	...	...	3
No of coils:									
Pre 1980 models	...	....	...	...	...	...	7.5
Post 1980 models	...	...	...	...	...	9.5
Free length:									
Pre 1980 models	...	...	...	...	...	...	43.50 mm (1.75 in)
Post 1980 models	...	...	...	...	...	46.04 mm (1.81 in)

Operating mechanism

Ball diameter	...	...	...	...	...	...	9.525 mm (0.375 in)
Conical spring	...	...	...	...	...	...	
No of working coils	...	...	...	...	...	2
Free length	...	...	...	...	...	...	10.30 mm (0.407 in)
Clutch operating rod:								
Diameter	....	...	...	...	...	...	5.60 mm (0.219 in)
Length	...	...	...	....	...	...	...	297.48 - 297.74 mm (11.712 - 11.722 in)

Primary chain

No of links	...	...	...	...	...	...	...	84
Pitch	...	...	...	...	...	...	...	9.525 mm (3/8 in)
Width	...	...	...	...	...	...	...	6.350 mm (¼ in)
Type	...	...	...	...	...	...	...	Triplex endless

Torque wrench setting

| | | | | | | lbf ft | kgf m |
| Clutch centre nut | ... | ... | ... | ... | ... | 70 | 9.66 |

Specifications relating to Chapter 4

Carburettor(s)									TR65	TR65T	TR7V	TR7	TR7T
Type	...	...	...	...	...	...	...	...	Amal Mk I	Amal Mk I	Amal Mk I	Amal Mk I	Amal Mk I
ID no	...	...	...	...	...	...	...	...	930/108	930/108	930/94	930/94	930/108
Nominal bore size	...	...	...	...	...	...	30 mm	30 mm	30 mm	30 mm	30 mm
Main jet	...	...	...	...	...	...	...	240	220	270	270	240

Needle:					
Jet size	.106	.106	.106	.106	.106
Type	STD	STD	STD	STD	STD
Position	1	1	2	1	1
Throttle valve cutaway	3	3	3.5	3	3

	T140V	T140E pre 1980	T140E post 1980 and T140D	T140ES, LE and EX	TSX
Type	Amal Mk I	Amal Mk II	Amal Mk II	Bing CV	N/av
ID no	930/92 and 930/93	2930/1 and 2930/2	930/9 and 930/8	64	
Nominal bore size	30 mm	30 mm	30 mm	32 mm	
Main jet	190	200	200	145	
Pilot jet	N/Av	25	20	45	
Starter jet	N/Av	50	35	N/Av	
Piston	N/App	N/App	N/App	5(UK) 3(US)	
Needle:					
Jet size	.106	.105	.106	2.66(UK) 2.64(US)	
Type	STD	2C3	2C3	STD	
Position	1	2	2	3	
Throttle valve cutaway	3	3	3	N/App	

Air cleaner type   Surgical gauze and metal gauze

Fuel tank capacity
Large tank   18.2 litres (4.0/4.8 Imp/US gal)
Small tank   13.6 litres (3.0/3.6 Imp/US gal)

Fuel octane rating (minimum)   97 Premium grade

Oil tank capacity   2.3 litres (4.0/4.8 Imp/Us pints)

Oil pump (2 and 4-valve types)
Plunger diameter:
    Feed   10.308 - 10.316 mm (0.4058 - 0.4061 in)
    Scavenge   12.367 - 12.375 mm (0.4869 - 0.4872 in)
Bore diameter:
    Feed   10.319 - 10.331 mm (0.4062 - 0.4067 in)
    Scavenge   12.375 - 12.388 mm (0.4872 - 0.4877 in)
Drive block:
    Width   12.624 - 12.649 mm (0.497 - 0.498 in)
    Clearance in plunger head   0.965 - 0.114 mm (0.0015 - 0.0045 in)
Valve spring free length   12.70 - 13.21 mm (0.500 - 0.520 in)
Valve ball diameter   5.556 mm (0.219 in)

Oil pressure release valve
Piston diameter   14.237 - 14.249 mm (0.5605 - 0.5610 in)
Body bore diameter   14.275 - 14.286 mm (0.5620 - 0.5625 in)
Spring:
    Free length   34.925 mm (1.375 in)
    No of coils   13
Operating pressure   60 psi (4.22 kg/cm^2)

Oil pressure switch operating pressure   3.0 - 5.0 psi (0.21 - 0.35 kg/cm^2)

Oil pressure readings
Idling   20 - 25 psi (1.41 - 1.75 kg/cm^2)
    Normal running:
        Pre EDA 30 000   65 - 80 psi (4.55 - 5.60 kg/cm^2)
        Post EDA 30 000   60 - 65 psi (4.22 - 4.55 kg/cm^2)

Torque wrench settings	lbf ft	kgf m
Oil pump nuts	5.0	0.7
Pressure release valve	20	2.8

Specifications relating to Chapter 5

Ignition system
Timing
Crankshaft position (BTDC):
    Fully advanced   ...   .   38°
Piston position (BTDC):
    Fully advanced   ...   .   10.4 mm (0.415 in)
Advance range:
    Contact breaker system   ...   .   24°
    Electronic system   ...   .   38°
Engine speed at full advance:
    Contact breaker system   ...   .   2500 rpm
    Electronic system   ...   .   3500 rpm

Contact breaker gap   ...   .   0.35 - 0.40 mm (0.014 - 0.016 in)

Spark plugs
Type:
    TR7V, T140V models   ...   .   Champion N3
    All other models   ...   .   Champion N5
Gap   ...   .   0.635 mm (0.025 in)
Thread size   ...   .   14 mm x 19.05 mm (0.75 in) reach

Ignition coil type
TR7V and T140V (+ earth)
and TR65 models   ...   .   Lucas 17M12
All other models...   ...   .   Lucas 17M6

Specifications relating to Chapter 6

Front forks
Type   ...   .   Telescopic, hydraulically damped
Oil capacity per leg   ...   .   190 cc (6.7/6.4 Imp/US fl oz)
Fork stanchion diameter   ...   .   34.556 - 34.569 mm (1.3605 - 1.3610 in)
Fork lower leg bore diameter   ...   .   34.620 - 34.671 mm (1.363 - 1.365 in)
Spring:
    No of coils   68
    Free length   ...   .   485.14 mm (19.10 in)

Steering head bearings
Type   ...   .   Tinken, taper roller
Outer diameter   ....   45.24 - 45.26 mm (1.781 - 1.782 in)
Bore   ...   .   19.05 - 19.07 mm (0.7500 - 0.7508 in)

Swinging arm
Bushes:
    Bore diameter   ...   .   25.40 - 25.44 mm (1.0 - 1.0015 in)
    Outer diameter   ...   .   28.56 - 28.58 mm (1.1245 - 1.1255 in)
    Length   ...   .   19.05 - 19.17 mm (0.750 - 0.755 in)
    Material type   ...   .   Pre-sized phosphor bronze
Sleeves:
    Outer diameter   ...   .   25.33 - 25.36 mm (0.9972 - 0.9984 in)
    Length   ...   .   55.75 - 55.88 mm (2.195 - 2.200 in)

Rear suspension units

	T140V and TR7V	**T140E, D, ES and EX and TR7**	**TR7T and TR65T**
Make	Girling	Girling	Marzocci
Type	Coil spring hydraulic damper unit	Gas-filled	Standard
Spring free length	241.30 mm (9.50 in)	241.30 mm (9.50 in)	228.60 mm (9.0 in)
Spring wire diameter	N/Av	N/Av	7.59 mm (0.29 in)

	TR65	**T140LE Royal**	**TSX**
Make	Girling	Marzocci	Paoli (or SW for US models)
Type	Gas-filled	Air assisted Strada	N/Av
Spring free length	228.60 mm (9.0 in)	210.0 mm (8.27 in)	N/Av

Spring wire diameter	6.80 mm (0.268 in)	7.20 mm (0.283 in)	N/Av
Air charge pressure	N/App	28 psi (1.96 kg/cm^2)	N/App

Torque wrench settings

	lbf ft	kgf m
Fork stanchion top bolts	30	4.14
Fork yoke pinch bolts	25	3.45
Front wheel spindle cap retaining nuts	15	2.07
Fork damper retaining plug	30	4.14
Fork damper retaining screw	18	2.49
Headlamp pivot bolts	10	1.4
Swinging arm pivot bolt retaining nut...	60	8.29

Specifications relating to Chapter 7

Tyres

	T140V, early TR7V and TR65	T140E, late TR7V and TR7	TR7T and TR65T
Make	Dunlop	Avon/Dunlop	Avon
Type	K70	Roadrunner/TT100	Mudplugger
Size:			
Front	3.25 x 19	4.10 x 19	2.0 x 21
Rear	4.0 x 18	4.10 x 18	4.0 x 18
Pressure:			
Front	24 psi (1.7 kg/cm^2)	28 psi (1.9 kg/cm^2)	22 psi (1.5 kg/cm^2)
Rear	24 psi (1.7 kg/cm^2)	32 psi (2.2 kg/cm^2)	24 psi (1.7 kg/cm^2)

	T140D, ES, LE and EX	TSX (US)	TSX (UK)
Make	Avon/Dunlop	Goodyear	Avon
Type	Roadrunner/TT100	Eagle	Speedmaster Mk II (front) Roadrunner (rear)
Size:			
Front	4.10 x 19	MJ90 x 19 Rib	3.25H 19
Rear	4.25 x 18	MT90 x 16 Low profile	5.10H 16
Pressure:			
Front	28 psi (1.9 kg/cm^2)		
Rear	32 psi (2.2 kg/cm^2)		

Wheels

	T140V, E, ES and and EX. TR7V, TR7 and TR65	TR65T and TR7T	T140E, D, ES and LE, TSX
Type	Conventional, steel spoked with chromed steel rim		Cast alloy.Lester or Morris (L or M)
Rim size:			
Front	WM2 x 19	WM1 x 21	19 x 2.15
Rear	WM3 x 18	WM3 x 18	18 x 2.50
Bearing size:			
Front:			
Right-hand	14 x 20 x 47 mm (0.55 x 0.79 x 1.85 in)	14 x 20 x 47 mm (0.55 x 0.79 x 1.85 in)	14 x 20 x 47 mm (0.55 x 0.79 x 1.85 in)
Left-hand	15 x 25 x 52 mm (0.59 x 0.98 x 2.05 in)	15 x 25 x 52 mm (0.59 x 0.98 x 2.05 in)	L - 14 x 20 x 47 mm (0.55 x 0.79 x 1.85 in) M - 15 x 25 x 52 mm (0.59 x 0.98 x 2.05 in)
Rear:			
Right-hand	15 x 25 x 52 mm (0.59 x 0.98 x 2.05 in) - except TR65	14 x 20 x 47 mm (0.55 x 0.79 x 1.85 in) - including TR65	L - 14 x 20 x 47 mm (0.55 x 0.79 x 1.85 in) M - 15 x 25 x 52 mm (0.59 x 0.98 x 2.05 in)
Left-hand	15 x 25 x 52 mm (0.59 x 0.98 x 2.05 in)	15 x 25 x 52 mm (0.59 x 0.98 x 2.05 in)	15 x 25 x 52 mm (0.59 x 0.98 x 2.05 in)

Spoke size:

Front, inner, right and left-hand	20 off 10 SWG, 196.85 mm (7.75 in) mean length, 96° head	20 off 10 SWG, 224.79 mm (8.85 in) mean length, 96° head	N/App
Front, outer, right and left-hand	20 off 10 SWG, 199.39 mm (7.85 in) mean length, 80° head	20 off 10 SWG 223.52 mm (8.80 in) mean length, 80° head	N/App
Rear, outer, left-hand	10 off 9 SWG 147.32 mm (5.8 in) mean length, 100° head	10 off 9 SWG, 147.32 mm (5.8 in) mean length, 100° head	N/App
Rear, inner, left-hand	10 off 9 SWG, 144.79 mm (5.7 in) mean length, 102° head	10 off 9 SWG, 144.79 mm (5.7 in) mean length, 102° head	N/App
Rear, inner and outer, right-hand	20 off 9 SWG, 182.80 mm (7.2 in) mean length, 135° head	20 off 9 SWG, 182.80 mm (7.2 in) mean length, 135° head	N/App

Rear wheel sprocket
No of teeth:
  Pre EDA 30 000 T140 and TR7 models   47
  Post EDA 30 000 T140 and TR7 models   45
  TR65, TR65T and TR7T models   47

Final drive chain
Pitch .. ..   15.875 mm (5/8 in)
Width   ...   .   9.525 mm (3/8 in)
No of links:
  Pre EDA 30000   107
  Post EDA 30000   106

Brakes
Front (all models):
  Type ...   .   Disc, hydraulically operated
  Disc diameter:
    Chromed ...   .   254.0 mm (10.0 in)
    Unchromed   ...   .   248.92 mm (9.80 in)
  Friction pads:
    Type   ...   .   DON 230 FGD08K - Pre EDA 30000
                                                     Dunlop sintered - Post EDA 30 000
    Minimum permissible lining thickness   ...   ...   1.98 mm (0.078 in)
Rear (T140 and TR7 pre engine no HN62501, TR65, TR65T and TR7T models):
  Type ....   Mechanically operated, internally expanding, single leading shoe, drum
  Drum diameter   ...   .   177.80 + 0.0508 mm (7.0 + 0.002 in)
  Minimum permissible shoe lining thickness   ...   ...   1.016 mm (0.040 in)
Rear (all other models) ...   .   Hydraulically operated disc brake of similar specifications to front brake

Speedometer drive
Drive gearbox ratio   ...   .   1.25 : 1
Drive cable length:
  Inner:
    Pre 1980 ...   .   171.7 cm (67.63 in)
    Post 1980...   ...   .   184.8 cm (72.75 in)
  Outer:
    Pre 1980 ...   .   167.6 cm (66.0 in)
    Post 1980   ...   .   180.9 cm (71.25 in)
Note: *Speedometer gearboxes fitted to pre 1980 models are fitted to right-hand side of rear wheel hub, whereas gearboxes fitted to post 1980 models are fitted to left-hand side of hub.*

Torque wrench settings	lbf ft	kgf m
Brake disc retaining bolt nuts	27	3.73
Brake drum retaining bolt nuts	15	2.10
Drum brake cam spindle retaining nut	20	2.80
Rear wheel spindle retaining nut	60	8.29

Specifications relating to Chapter 8

Electrical equipment	T140V and TR7V (+ earth)	1979/80 T140 and TR7 (kickstart)	T140, TR7 and TR7T 1980/81 electric start and 1981 kickstart	1981 TR65
Battery:				
Capacity	8 Ah	9 Ah	14 Ah (E/S) 9 Ah (K/S)	9 Ah
Voltage	12 volt	12 volt	12 volt	12 volt
Earth	Positive	Negative	Negative	Negative
Rectifier	2DS 506	3DS	3DS	3DS
Alternator	RM21	RM24	RM24	RM21
Zener diode	ZD715	ZD715 Single	XD715A Triple pack	ZD715 Twin pack
Ignition coil:				
Type	17M12	17M6	17M6	17M12
Voltage	12 volt	6 volt	6 volt	12 volt
Ignition switch	S45	149SA	149SA	149SA
Ignition pick-up assembly	N/App	5PU	5PU	N/App
Ignition amplifier	N/App	AB11	AB11	N/App
Condenser	54420128	N/App	N/App	54420128
Contact breaker	6CA	N/App	N/App	10CA
Auto advance unit	6CA	N/App	N/App	10CA
Horn	6H	6H	6H	6H
Flasher unit	8FL	8FL	8FL	8FL
Handlebar switch	169SA	N/Av	N/Av	N/Av

Note: *All items of electrical equipment are of Lucas manufacture*

Bulbs	Lucas number	Type (12 volt)
Headlamp:		
Pre 1981 T140 and TR7 models. TR65T and TR7T models	410	45/40W pre-focus
Post 1981 T140 and TR7 models. TR65 model:		
UK	N/Av	60/45W sealed beam. LH Dip
US	N/Av	60/50W sealed beam. RH Dip
Pilot lamp:		
Pre 1981 T140 and TR7 models. TR65T and TR7T models	989	6W M.cc
Post 1981 T140 and TR7 models. TR65 model	233	6W
Stop/tail lamp:		
TR65T and TR7T models	382	5/21W
All other models	380	5/21W
Warning lights	281	2W
Speedometer light	504	3W
Tachometer light	504	3W
Direction indicators	382	21W

Fuse rating	35 amp

Torque wrench settings	lbf ft	kgf m
Alternator rotor retaining nut	40	4.1
Alternator stator retaining nuts	20	2.8
Zener diode retaining nut	1.5	0.21
Rectifier retaining nut	1.5	0.21

Dimensions and weight

Wheelbase	142 cm (56 in)
Overall length	222 cm (87.5 in)
Overall width:	
UK and general export	73.5 cm (29 in)
USA	84 cm (33 in)
Ground clearance	18 cm (7 in)
Seat height:	
TR65T and TR7T	82.5 cm (32.5 in)
TSX	76 cm (30 in)
All other models	79 cm (31 in)

Dry weight:

TR65	179.3 kg (395 lb)
TR65T and TR7T	173.9 kg (383 lb)
T140 EX	218.8 kg (482 lb)
TR7 and all other T140:	
Kickstart models	179.3 kg (395 lb)
Electric start models	194.8 kg (429 lb)
TSX	188.4 kg (415 lb)

Recommended lubricants

Component	Lubricant	Quantity
Engine	SAE 20W/50 engine oil	2.27 lit (4.0/4.8 Imp/US pint)
Gearbox	SAE 90 EP gearbox oil	500 cc (0.9/1.0 Imp/US pint)
Primary chaincase	SAE 20W/50 engine oil	150 cc (5.28/5.07 Imp/US fl oz) Initial fill
Front forks	Automatic transmission fluid (ATF)	190 cc (6.68/6.42 Imp/US fl oz) each leg
Front brake hydraulic system	SAE J 1703 (UK) or DOT 3 (US) hydraulic fluid	189 cc (6.65/6.39 Imp/US fl oz) approximate
Rear brake hydraulic system	SAE J 1703 (UK) or DOT 3 (US) hydraulic fluid	205 cc (7.22/6.93 Imp/US fl oz) approximate
Steering head bearings	Lithium base high melting point grease	As required
Swinging arm bearings	Lithium base high melting point grease	As required
Wheel bearings (unsealed)	Lithium base high melting point grease	As required
Brake pedal link pivot	SAE 10W/30	As required
Brake pedal spindle	SAE 10W/30	As required
Control cables (non-nylon)	SAE 10W/30	As required

1 Introduction

Since manufacture of the Triumph 750cc vertical twins passed into the hands of the Meriden Motor Cycle Co-operative, a certain amount of design updating has been carried out. This has been dictated primarily by American motorcycle legislation relating to emission control and construction and use which, if not observed, would automatically debar the continued export-ation of the existing models. Styling changes have also been nec-essary to give the machines a more modern image in order to make them competitive with the various machines made by foreign manufacturers.

Although the basic design of the 750cc vertical twin still adheres to the 'tried and proven' formula that can be traced back many years, some new features have been included to make the machine generally more efficient.

In August of 1975, the Bonneville T140V and Tiger TR7V models were re-introduced with a left-foot gearchange and a rear 10 inch disc brake. Further improvements in the form of electronic ignition, a negative earth electrical system, Veglia type instruments, Amal Mk II carburettors and a redesigned (parallel tract) cylinder head were introduced in September of 1978, these improvements being incorporated on the T140E and TR7 (single Amal 930 carburettor) models.

Up to the time of writing this Chapter (May of 1983) styling changes and design improvements on the Bonneville models have included the incorporation of an electric starter system, the fitting of Bing Constant Velocity carburettors and further variations in seat, petrol tank and side panel designs. The basic electric start Bonneville is designated the T140 ES. Later varia-tions of this machine are the T140 LE Royal, the T140 EX Executive and the TSX. Whereas the Royal and TSX differ from standard mainly in cosmetic design, to appeal to certain sections of the buying public, the Executive goes one step further in having a full range of high-quality touring equipment fitted.

Until recently, Tiger models have been marketed in the UK as a single carburettor version of the Bonneville only, the two models being developed together. Seeing the recent demand for a large capacity on/off road machine, Triumph have been quick to transform the Tiger from its rather staid guise into a machine which is not only visually striking but which can operate with reasonable efficiency over most types of terrain. This machine is known as the Tiger Trail and is designated the TR7T.

With the Bonneville having lost some of the sporting image for so long associated with its name, Triumph have moved to produce a light sporting machine of economic design, the TR65 Thunderbird. This they have done by taking the standard 750cc engine unit and shortening its stroke to reduce the capacity to 650cc. Fed by a single Amal Mk I carburettor, this engine has surprised many motorcyclists by its ability to produce almost as much usable power as the 750cc unit whilst producing much less vibration yet being able to operated in a higher rev range.

Although many of the major cycle components of the Thunderbird are identical to those fitted to the Bonneville, economic measures taken during its design are apparent in the incorporation of a re-designed instrument console (UK only), a rear drum brake, contact breaker ignition and a siamese exhaust system. No prop stand is fitted as standard and of course, no electric starter. To produce a more flexible Trail machine, the 650cc engine unit has now been incorporated into a cycle framework similar to that used for the TR7T, the resulting machine being designated the TR65T.

On all models throughout the Triumph range covered by this Chapter, different styling effects have been obtained by fitting petrol tanks, side panels and seats of various designs. As well as the standard twin-downpipe exhaust system, siamese systems of both high- and low-level designs have been introduced and equipped with the latest type of silencer which succeeds in muting the exhaust note without appreciably affecting engine performance. Although great use is still being made of the standard wire-spoked wheel, the fitting of cast alloy wheels is now optional on most model types, their use adding greatly to appearance. To complete the improvement in engine specifi-cation, a new four-valve oil pump is now fitted to all engine types. Added to this, Triumph are constantly seeking to improve the already excellent handling qualities of their machines by fitting various types of rear suspension unit.

Although the model types covered by this Chapter are similar in a great many respects to the models covered in the first eight Chapters of this manual, reference should always be made to this Chapter first in view of the need to follow a modified procedure or use different settings, when certain components have to be removed and replaced. Where no information is given in this Chapter, it can be assumed that the procedure is identical to that described for the earlier models.

Certain changes have been made in the Routine Maintenance procedures and it is important that attention is paid to the relevant details set out in this Chapter.

It should be noted that the T140V and TR7V models prior to the introduction of those fitted with the left-hand gearchange pedal and hydraulic disc rear brake are covered in Chapters 1 to 8.

2 Frame number location

The frame identification number is now found on a plate attached to the top frame tube, on the left-hand side, immediately to the rear of the steering head. The plate also confirms that the machine conforms to certain motor vehicle safety standards. It is also stamped on the left-hand front down tube, close to the steering head.

3 Routine maintenance

Weekly, or every 250 miles (400 km)

1 Safety inspection
Give the complete machine a close and thorough visual inspection, checking for loose nuts, bolts and fittings, frayed control cables, damaged brake hoses, severe oil and petrol leaks, etc.

2 Legal inspection
Check the operation of the electrical system, ensuring that the lights and horn are working properly and that the lenses are clean. Note that in the UK it is an offence to use a vehicle on which the lights are defective. This applies even when the machine is used during daylight hours. The horn is also a statutory requirement.

3 Tyre pressure check
Check the tyre pressures. Always check with the tyres cold, using a pressure gauge known to be accurate. It is recommended

that a pocket pressure gauge is purchased to offset any fluctuation between garage forecourt instruments. The tyre pressures for each model type are given in the Specifications Section of this Chapter.

At this juncture also inspect the actual condition of the tyres, ensuring there are no splits or cracks which may develop into serious problems. Also remove any small stones or other small objects of road debris which may be lodged between the tread blocks. A small flat-bladed screwdriver will be admirable for this job. Examine the amount of tread remaining on the tyre. No tyre should be worn beyond the relevant legal limit.

4 Final drive chain lubrication
In order that the life of the final drive chain be extended as much as possible, regular lubrication is essential. Intermediate lubrication should take place with the chain in position on the machine. The chain should be lubricated by the application of one of the proprietary chain greases contained in an aerosol can. Ordinary engine oil can be used, though owing to the speed with which it is flung off the rotating chain, its effective life is limited.

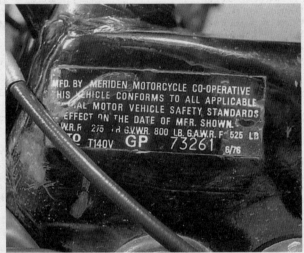

Frame number location

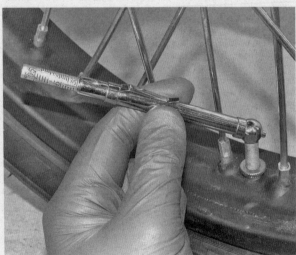

Check tyre pressures with an accurate gauge

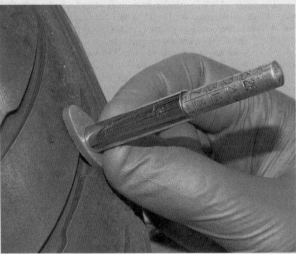

Check tyre tread depth

5 Engine oil level check

The oil level in the oil-in-frame reservoir should be checked, and if necessary, topped up with SAE 20W/50 engine oil. The filler cap, found close to the nose of the dualseat, has an integral dipstick for this purpose. There is no necessity to check the level of oil in the primary chaincase, as this is topped up from the engine breather.

Before refitting the filler cap, start the engine and check that oil is returning to the reservoir.

6 Battery electrolyte level check

Do not omit to check the electrolyte level of the battery, or serious damage will occur, which will greatly shorten the working life of the battery. The electrolyte level should lie between the two lines marked on the battery case. If the level is low, top it up with distilled water only, unless it is known that spillage of acid has occurred. In this latter case, it is necessary to remove all traces of acid, which is highly corrosive, by washing all the affected parts with copious amounts of water, or water in which a neutralising agent, such as baking soda, has been dissolved. Top up the battery with acid of the correct specific gravity (1.260 - 1.280) under these circumstances. Take care not to overfill the battery.

Oil filler cap has integral dipstick

Location of battery and rear brake reservoir

7 Disc brake hydraulic fluid level check

Place the machine on its main stand on an area of flat and level ground. Move the handlebars to full right-hand lock to position correctly the fluid reservoir of the front brake master cylinder. Where applicable, expose the fluid reservoir of the rear brake master cylinder by detaching the right-hand sidepanel from the machine or by lifting the seat. To check the fluid level in each reservoir, carefully unscrew the reservoir cap and carefully lift out the diaphragm whilst taking care not to allow any brake fluid to drip onto component parts of the machine. Note that hydraulic fluid will damage paintwork and plastic component parts and should therefore be wiped up as soon as a spillage occurs. A level mark which runs around the inside of the reservoir, approximately 6 mm (0.25 in) from its top, provides an indication as to the correct fluid level.

If the fluid level in the reservoir is seen to be excessively low, then suspect a fluid leakage in the brake system and carry out a thorough examination of the system to detect any faults before replenishing the reservoir. Replenish the reservoir with an hydraulic fluid of SAE J 1703 (UK) or DOT 3 (US) specification; no other specification of fluid should be used as an incorrect fluid may perish the rubber seals within the brake system thereby causing brake failure. Check that the reservoir diaphragm is not split or perished before pushing it carefully into the reservoir and check that the sealing washer of the cap is not damaged before refitting the cap and tightening it, hand-tight. The breather hole in the reservoir cap must be kept clear.

Monthly, or every 1000 miles (1600 km)

1 Control cable lubrication

Lubricate each control cable thoroughly with a light oil (SAE 10W/30). If nylon lined replacement cables have been fitted the cables should not be oiled. Should oil be introduced into a nylon lined cable, it may cause the nylon to swell thereby causing total cable seizure.

Proprietary cable oilers are available for oiling the cables whilst they are in position. The alternative is to make up a plasticine funnel as shown in the accompanying illustration, and permit the oil to seep down the length of the cable. This latter method, although simpler, has the disadvantage that the cable needs to be detached first.

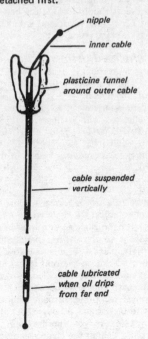

Fig. 9.1 Control cable lubrication (non-nylon)

2 Swinging arm fork pivot lubrication

A grease gun filled with grease of the lithium-based, high melting point type, should be applied to the grease nipples on the pivot tubes of the swinging arm fork and grease pumped into the pivot bearings until clean grease can be seen to exude from either of their ends. On completion of this operation, wipe away all excess grease.

Swinging arm pivot is equipped with grease nipples

3 General lubrication

Work around the machine, applying grease or oil to any pivot points. These points should include the handlebar lever pivots, the rear brake pedal link pivot and spindle, the centre stand pivots and the prop stand pivot.

4 Primary chain adjustment check

The primary drive chain is of the triplex type, and it is tensioned by means of a slipper type tensioner located below the bottom run of the chain. If the top inspection cap of the primary chaincase is unscrewed and removed, the tension of the chain can be checked with a finger. THIS CHECK SHOULD NEVER BE MADE WITH THE ENGINE RUNNING, OR WHILST THE ENGINE IS TURNED OVER BY THE STARTER. The correct amount of slack in the chain should be 9.5 mm (0.4 in).

Pre-1980 models

With the type of chain tensioner mechanism fitted to these models, in order to adjust the chain tension, it may be necessary to remove the left-hand silencer to gain access to the hexagonal pillar bolt adjacent to the centre stand left-hand lug. Before removing this bolt, place a receptacle below it to catch any oil that may drain off. Remove the bolt to expose the slotted head adjuster. To take up slack in the chain, turn the adjuster clockwise, or to increase the amount of slack, anti-clockwise. A flexible screwdriver will prove virtually essential, if the inaccessible adjuster is to be reached. Always check the chain tension in several different places by rotating the engine. When adjustment is correct, refit and tighten the chaincase plug and the left-hand silencer. Replenish the chaincase so that it contains 150 cc (5.28/5.07 Imp/US fl oz) of SAE 20W/50 engine oil and then refit the inspection cap.

Post-1980 models

With the type of chain tensioner mechanism fitted to these models, it is first necessary to remove the left-hand footrest from the machine. Removal of the footrest will permit rotation of spanners fitted over the adjuster bolt and its locknut. To adjust the chaintension, hold the adjuster bolt steady and slacken its locknut. Move the locknut clear of the chaincase plug through which the adjuster bolt passes and then rotate the adjuster bolt to adjust the chain tension whilst using a spanner to hold the plug steady. Take care not to unscrew the chaincase plug as this will permit oil to drain from the chaincase. Turning the adjuster bolt clockwise (inwards) will take up slack in the chain. Remember to check the chain tension at several points along its length, this will show any tight point in the chain. Upon completion of adjustment, secure the adjuster bolt in position by tightening its locknut and then refit the footrest.

Use this opportunity to check the level of oil in the primary chaincase. If necessary, replenish the chaincase so that it contains 150cc (5.28/5.07 Imp/US fl oz) of SAE 20W/50 engine oil before refitting the inspection cap.

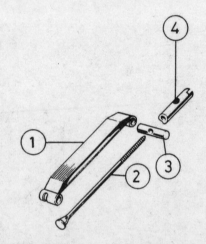

Fig. 9.2 Pre 1980 primary chain tensioner assembly

1 Tensioner blade   3 Tensioner
2 Rod   4 Adjuster

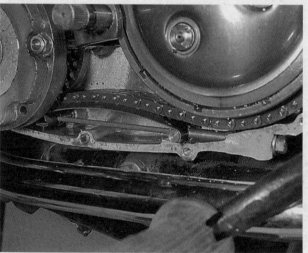

Primary chain tensioner is in base of chaincase

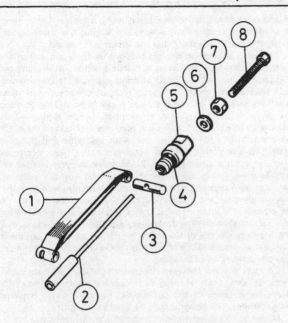

Fig. 9.3 Post 1980 primary chain tensioner assembly

1 Tensioner blade	5 Plug
2 Rod	6 Washer
3 Tensioner	7 Locknut
4 O-ring	8 Bolt

5 Final drive chain adjustment check

Adjustment of the final drive chain is achieved by means of drawbolts which are threaded into brackets which fit over each end of the rear wheel spindle. When in correct adjustment, the final drive chain should have 19 mm (0.75 in) of free movement with the rider seated on the machine, or 44 mm (1.75 in) with the machine placed on its centre stand. Any measurement of free movement must be made at the mid-point of the chain lower run and several measurements should be made whilst rotating the rear wheel in order to determine the tightest point in the chain.

If the chain is too slack, then the wheel spindle retaining nut should be loosened and the wheel drawn rearwards by means of the drawbolts until chain tension is correct. Before attempting to rotate each of the drawbolts, slacken their locknuts. Where the machine is fitted with a rear drum brake, loosen the nut which retains the torque rod to the brake backplate. Make sure that both drawbolts are rotated an equal amount or the wheels will become out of track.

Nip tight the wheel spindle retaining nut; on drum brake models, apply the brake pedal to centralise the brake backplate before doing this. Recheck the chain adjustment and if there is any doubt about the wheel alignment, check by placing a straight plank of wood parallel to the machine, so that it touches both walls of the rear tyre. If wheel alignment is correct, it should be equidistant from either side of the front wheel tyre, when tested on both sides of the rear wheel; it will not touch the front tyre because this tyre has a smaller cross section.

Finally, tighten the wheel spindle retaining nut to the specified torque loading, lock the drawbolts in position by tightening their locknuts and, on drum brake models, retighten the torque rod retaining nut. Before taking the machine onto the road, roll the machine forward and apply the rear brake to ensure that it is functioning correctly.

6 Final drive chain lubrication

Although a satisfactory result can be achieved without removing the chain from the machine by using one of the special chain lubricant aerosols, it is better to remove the chain at regular intervals for more detailed attention. Refer to Section 14 of Chapter 7 for the relevant details.

Six weekly or every 1500 miles (2400 km)

1 Engine oil change

The oil-in-frame reservoir holds approximately 2.27 lit (4.0/4.8 Imp/US pints) of SAE 20W/50 engine oil, so it is necessary to place a receptacle capable of holding more than this amount under the drain plug at the base of the frame reservoir. It is preferable to drain the oil whilst it is hot, so that it will be more fluid and thus drain off much more easily. Remove the filler cap, and the hexagon headed drain plug from the centre of the reservoir base plate and allow the oil to drain. Whilst the oil is draining, check the sealing washer beneath the head of the drain plug for damage or deterioration and renew it if necessary.

On completion of the oil draining, remove the four nuts with spring washers that retain the reservoir base plate so that it too can be detached and the gauze filter cleaned with petrol. Any stubborn traces of contamination blocking the gauze can be removed by brushing the affected area with a small brush soaked in petrol; a used toothbrush is ideal. Remember to take the necessary fire precautions when carrying out this cleaning procedure and always wear eye protection against any petrol that may spray back from the brush. On completion of cleaning, closely inspect the gauze area for any splits or holes that will allow the passage of sediment through it and into the oil system. Renew the gauze filter if it is in any way defective. Flush out the reservoir with new flushing oil, then refit the base plate, not forgetting the two gaskets which fit each side of the filter flange. If damaged, they should be renewed, to prevent a permanent oil leak occurring. Check also that the four spring washers have not become flattened. If this is the case, then renew them as they will have lost their locking function.

Tighten the base plate retaining nuts, evenly and in a diagonal sequence so as to avoid distortion of the plate, and then refit the drain plug with its serviceable sealing washer. Refill the reservoir with the specified amount of oil and allow time for it to settle before checking its level with the dipstick which is integral with the filler cap. Before refitting the filler cap, start the engine and allow it to idle for a few minutes until oil is seen to be returning to the reservoir.

Engine oil drain plug is at base of frame reservoir

Three monthly or every 3000 miles (4800 km)

1 Valve clearance check

To check the valve clearances, remove both spark plugs and the finned rocker box covers. Turn the engine over by means of the kickstart (or alternatively, by placing the machine on its centre stand and turning the rear wheel after top gear has been engaged) until one of the inlet rockers moves downwards, opening the valve. Continue turning until the valve is open fully, which will ensure the other inlet valve is seated correctly for the valve clearance to be measured. The gap should be 0.203 mm (0.008 in) on all models except the TR7T, where it should be 0.152 mm (0.006 in). Measure the gap with a feeler gauge of the required thickness.

If adjustment is necessary, slacken the locknut at the end of the rocker arm and turn the adjuster either clockwise or anti-clockwise, to either decrease or increase the clearance to the desired amount. Only a fraction of a turn is likely to be necessary. When the setting is correct, tighten the locknut, making sure the adjuster does not turn by holding it with a spanner, and recheck the clearance. If correct, turn the engine over until the valve just checked is fully open, then repeat the adjustment procedure for the other valve.

A similar procedure should be adopted for the exhaust valve, positioning one fully open whilst the second is checked. The recommended clearance in this instance is 0.152 mm (0.006 in) for all models.

Before refitting the finned rocker box covers, check that the sealing gaskets are in good condition. Refit the covers, making sure the retaining bolts and screws are tightened evenly.

It is important to note that valve clearances should always be checked when the engine is COLD.

2 Spark plug check

Whilst the spark plugs are removed for checking the valve clearances, it is recommended that they be cleaned and checked as follows.

Triumph fit a Champion N5 spark plug as standard equipment to all of the machines covered in this Chapter (except the T140V and TR7V models which are equipped with a Champion N3). The recommended gap between the plug electrodes is 0.635 mm (0.025 in). Each plug should be cleaned and the gap checked and reset at regular service intervals. In addition, in the event of a roadside breakdown, where the engine has mysteriously 'died', the spark plug should be the first item checked.

The plug should be cleaned thoroughly by using one of the following methods. The most efficient method of cleaning the electrodes is by using a bead blasting machine. It is quite possible that a local garage or motorcycle dealer has one of these machines installed on the premises and will be willing to clean any plugs fr a nominal fee. Remember, before fitting a plug cleaned by this method, to ensure that there is none of the blasting medium left between the porcelain insulator and the plug body. An alternative method of cleaning the plug electrodes is to use a small brass-wire brush. Most motorcycle dealers sell such brushes which are designed specifically for this purpose. Any stubborn deposits of hard carbon may be removed by judicious scraping with a pocket knife. Take great care not to chip the porcelain insulator round the centre electrode whilst doing this. Ensure that the electrode faces are clean by passing a small fine file between them; alternatively, use emery paper but make sure that all traces of the abrasive material are removed from the plug on completion of cleaning.

To reset the gap between the plug electrodes, bend the outer electrode away from or closer to the central electrode and check that a feeler gauge of the correct size can be inserted between the electrodes. The gauge should be a light sliding fit.

Never bend the central electrode or the insulator will crack, causing engine damage if the particles fall in whilst the engine is running.

Always carry a spare spark plug of the correct type.

Beware of overtightening a spark plug, otherwise there is risk of stripping the threads from the aluminium alloy cylinder head. The plug should be sufficiently tight to seat firmly on its sealing washer, and no more. Use a spanner which is a good fit to prevent the spanner from slipping and breaking the insulator.

Before fitting each spark plug in the cylinder head, coat its threads sparingly with a graphited grease. This will prevent the plug from becoming seized in the head and therefore aid future removal.

When reconnecting the suppressor cap to each plug, make sure that the cap is a good, firm fit and is in good condition; renew its rubber seals (where fitted) if they are in any way damaged or perished. Later types of cap contain a suppressor which eliminates both radio and TV interference.

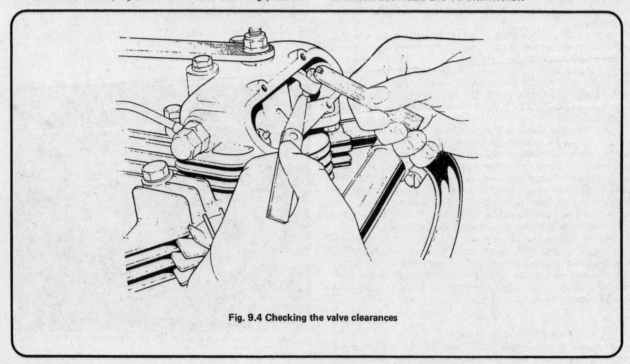

Fig. 9.4 Checking the valve clearances

3 Gearbox oil level check

Check the gearbox oil level by first placing the machine on its centre stand on an area of flat and level ground and then unscrewing the oil level plug (C) from the centre of the drain plug (D) located in the underside of the gearbox, on the right-hand side. A small amount of oil should drain out of the level tube.

Remove the filler plug (B) from the gearbox outer cover and slowly pour in SAE 90EP gearbox oil until fresh oil is seen to be dripping from the level plug orifice, then leave the machine until the dripping has stopped. Check the condition of the sealing washer and O-ring, renewing them if necessary, then refit the level and filler plugs and tighten them securely, but be careful not to overtighten them.

Note: Do not assume that the level is correct merely because oil drips out as soon as the level plug is removed; the correct level can be established only by carrying out the full procedure described above.

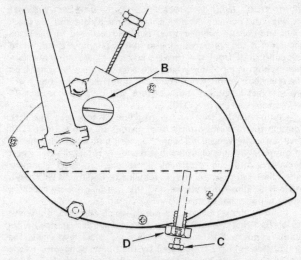

Fig. 9.5 Checking the gearbox oil level

B Filler cap   D Oil drain plug
C Oil level plug

4 Clutch adjustment check

It is recommended that the clutch operating mechanism is checked for correct adjustment at regular service intervals. This is a straightforward operation which should be carried out in accordance with the instructions given in Section 2 of Chapter 3 of this Manual. It is, however, interesting to note that whereas Triumph maintain that a clearance of 1.5 mm (0.06 in) should exist between the clutch operating rod and the adjuster screw in the clutch pressure plate, service information issued for later models (May 1982 on) states that rather than back off the adjuster screw one complete turn to obtain the necessary clearance, the adjuster screw should only be backed off one half turn.

5 Air filter element examination

The two air filter elements fitted to each machine are formed from a sandwich of corrugated metal gause and surgical gauze, this sandwich being bonded between two end plates which are rubber coated; this rubber coating forms an effective seal when each element is fixed to its housing.

To remove each element, first detach the side panels (where fitted) and then remove the single nut with lock washer which retains each filter housing cover in potition. Note the fitted position of each element before pulling it from its housing and carrying out the following examinations.

Carefully examine the gauze of the elements for signs of splitting or separation from its end plates. Any damage of this kind will necessitate renewal of the element. The same applies if the sealing area of any one end plate is found to be damaged or perished. If the element is seen to be sound, then clean the gauze by rinsing it in clean petrol or paraffin before carefully blowing it dry with a jet of compressed air. Remember to take the necessary precautions against fire whilst using petrol and take care to protect the eyes against any blow back of petrol from the air jet.

Relocate each filter element in its housing whilst making sure that it is fitted the correct way up, otherwise the air supply to the carburettor(s) will be shut off. Great care must be taken when positioning each element and its housing cover to ensure that no incoming air is allowed to bypass the elements. If this is allowed to happen, it will allow any dirt or dust that it normally retained by the element to find its way into the carburettor and crankcase assemblies; it will also effectively weaken the petrol/air mixture.

Note that if the machine is being run in a particularly dusty or moist atmosphere, then it is advisable to increase the frequency of cleaning each element. Never run the engine without both elements fitted. This is because each carburettor is specially jetted to compensate for the addition of this component and the resulting weak mixture will cause overheating of the engine with the probable risk of severe engine damage.

6 Carburettor examination (Amal type)

Each Amal Mk I or II Concentric has a drain plug located in the base of its float bowl. This plug is hollow and is designed to trap any sediment present in the petrol entering the float bowl. It is recommended that the drain plug be removed and cleaned at regular service intervals.

Relocate each air filter element correctly

7 Wheel bearing check

Severe wear in a wheel bearing will be indicated by a whine or rumble emitting from the bearing whilst the machine is in motion. To obviate the chances of any one bearing reaching such a severe state of wear that its breakdown is imminent, it is recommended that the following check be carried out at regular service intervals.

Position the machine on an area of flat and level ground and then place it on its centre stand. Position a block of wood or a similar support beneath the engine crankcase so that the front wheel is raised clear of the ground. Check that the machine is firmly supported so that it will not topple, grasp one lower fork leg with one hand and spin the wheel with the other. Vibration felt through the fork leg or any rumble emitting from the wheel hub will indicate bearing wear. Carry out a further check by allowing the wheel to rest on full lock whilst grasping the wheel with both hands placed 180° apart on the wheel rim. With the forks held steady, feel for any play in the bearing by attempting to move the wheel from side to side. If in the slightest doubt as to the condition of a bearing, then remove the wheel and investigate further by carrying out the instructions listed in Section 7 of Chapter 7.

The bearings of the rear wheel should be examined by using a method similar to that given for examination of the front wheel bearings. It will be appreciated, however, that it will not be possible to spin the wheel freely without first disconnecting the final drive chain from the rear wheel sprocket. If it is decided to disconnect the chain, then take care to prevent it coming into contact with the ground and thus becoming contaminated with dirt or grit. Take note of the information given in Section 14 of Chapter 7 before reconnecting the chain.

8 Disc brake pad wear check

It is most important that the pads of both the front and rear (where applicable) disc brakes are examined for wear at regular service intervals. The pads will require renewing when the pad lining thickness reaches a minimum of 2 mm (0.08 in). It is important not to confuse this with the overall thickness of the pad plus its metal backing, since if the pad wears down to the bare metal, the disc itself will be badly damaged. Worse still, the loss in braking efficiency may give rise to a serious accident.

To examine or renew the pads, first detach the chrome plated cover from around the caliper; this is retained by two screws. If the two split-pins that retain the brake pads are then removed, the pads can be lifted out, one at a time.

When refitting the existing pads or fitting new ones, push the actuating plungers of the caliper back with a screwdriver, otherwise there will not be sufficient clearance to reinsert the pads. Use new split-pins to retain the pads in place and do not omit to bend the ends of each pin over.

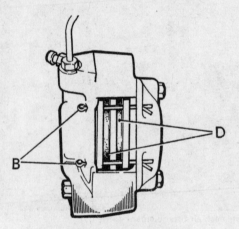

Fig. 9.6 The disc brake caliper

B Split-pins   D Brake pads

Six monthly or every 6000 miles (9600 km)

1 Contact breaker check

Remove both spark plugs and detach the small circular cover from the engine unit timing cover to expose the contact breaker assembly. Turn the engine over by means of the kickstart (or alternatively, by placing the machine on its centre stand and turning the rear wheel after top gear has been engaged) until the scribe line on the contact breaker cam is in exact alignment with the nylon heel of one set of contact breaker points. With the points thus fully apart, measure the gap between them by using a set of feeler gauges. If correct, the gap should be between the limits of 0.35 - 0.40 mm (0.014 - 0.016 in). If adjustment is necessary, then slacken the fixed contact securing screw and move the contact by means of the eccentric screw until the gap

is correct. Retighten the contact securing screw. Re-check the gap and if still correct, turn the engine until the second set of points align with the scribe mark in similar fashion. This set of points should then be checked, and if necessary, adjusted by following an identical precodure.

If the points are burnt or pitted, they should be removed from the contact breaker base plate and cleaned with fine emery cloth or an oilstone, before the gap is checked. Make sure the points are wiped with a cloth moistened with petrol first, so than any residual oil or abrasive does not act as an insulator. When refitting the points on their base plate, make sure the insulating washers are in their correct position, otherwise the whole assembly may be earthed permanently, isolating the ignition system. Refer to Chapter 5, Section 5, for further information. When the points are removed, it is helpful to place a few drops of clean engine oil on the automatic advance unit which lies behind the base plate, to prevent corrosion taking place.

2 Ignition timing check (contact breaker system)

With the contact breaker points correctly set and the spark plugs and contact breaker cover removed, proceed to check the accuracy of the ignition timing as follows. Remove the threaded inspection cap from the forward section of the primary chaincase; within is a pointer which will align with a timing mark on the alternator rotor. When the fixed pointer and the rotor mark are in exact alignment, both pistons will be positioned at 38° before top dead centre (TDC).

Commence by placing the machine in top gear and then rotating the rear wheel until both pistons are seen to be at TDC; they are easily viewed through the spark plug holes with the aid of a torch. With the pistons thus set, slowly rotate the rear wheel backwards until the rotor mark is brought into exact alignment with the fixed pointer. It now remains to determine which cylinder is on the firing stroke and this can be done by checking which cylinder has both of its valves closed.

Identify the set of contact points which operate the cylinder set on its firing stroke. The drive side cylinder is operated by the points connected to the black/white electrical lead whilst the timing side cylinder is operated by the points connected to the black/yellow lead. The relevant set of points should be just beginning to open; this can be checked by using a battery and bulb connected in circuit with the contact breakers as shown in the figure accompanying this text. Alternatively, a multimeter set on its resistance function or a 0.038 mm (0.0015 in) feeler gauge placed between the points can be used. Note that the contact breaker cam must be set in the fully advanced position during this check. The best method of achieving this is to use the automatic advance unit centre bolt as a holding tool. This can be done by first removing the bolt and then selecting a washer which will fit over the shank of the bolt and which has a diameter roughly equal to that of the contact breaker cam; the purpose of this washer is to act as a spacer between the bolt head and the cam so that once the cam is turned to its fully advanced position (fully clockwise) and the bolt refitted and tightened, the cam will remain locked in position.

If the points are not on the point of opening or the rotor mark is brought into alignment with the fixed pointer, then slacken the two pillar bolts which retain the contact breaker base plate in position and rotate it in the appropriate direction until the points are correctly set. Lock the base plate in position by tightening the pillar bolts and then rotate the crankshaft slowly through 360°. Check the second set of points for correct adjustment and if there is any discrepancy, slacken the secondary base plate retaining screws so that it can be moved independently of the main base plate moved previously. Use the eccentric adjuster screw to find the correct points position and retighten the secondary base plate retaining screws before rechecking the points setting.

Refit all disturbed engine components. Do not omit to remove from beneath the head of the advance unit centre bolt the washer used to lock the contact breaker cam in position, and lubricate the cam surface with Shell/Retinax A grease or an equivalent before refitting the cover to the assembly housing.

A stroboscopic lamp can, of course, be used to check the ignition timing should this item of equipment be available. This instrument provides by far the most accurate means of verifying the accuracy of the ignition setting and full instructions as to its use are given in Section 17 of this Chapter.

3 Spark plug renewal

Remove and discard the spark plugs, regardless of their condition. Although the plugs may give acceptable performance after they have reached this mileage, it is unlikely that they will still be working at peak efficiency.

The correct type of spark plug is a Champion N5 (except T140V and TR7V models which are fitted with Champion N3). Before fitting each new plug, adjust the gap between the electrodes to 0.635 mm (0.025 in), coat its threads sparingly with graphite grease and check that the aluminium crush washer is in place on the plug.

4 Gearbox oil change

Position the machine on an area of flat and level ground and then place it on its centre stand. Place a container of at least 500 cc (0.9/1.0 Imp/US pint) capacity beneath the gearbox level/drain plug. Remove the level plug from the centre of the drain plug and then unscrew and carefully withdraw the drain plug whilst taking care not to damage its level tube. Whilst waiting for the oil to drain, clean both the drain and level plugs and inspect their sealing washers for damage or deterioration, renewing each one as necessary.

Refit and tighten the drain plug, but NOT the level plug at this stage. Unscrew the gearbox filler cap, found close to the kickstart lever, and add sufficient SAE 90 EP oil until it commences to dribble from the level plug orifice. Allow the excess to drain off, then refit and tighten the level plug and also the filler plug. The gearbox should hold approximately 500 cc (0.9/ 1.0 Imp/US pint) of oil.

5 Front fork oil change

Position the machine on an area of flat and level ground and then place it on its centre stand. Place a container of at least 190 cc (6.7/6.4 Imp/US fl oz) capacity adjacent to the left and right-hand ends of the front wheel spindle. Remove both drain plugs from the lower fork legs and allow the oil contect to drain off. If desired, it can be expelled at a greater rate by compressing the forks several times.

Refit the drain plugs and tighten them, ensuring the sealing washers are in good condition. Remove the handlebars (where necessary), the large hexagonal (metal) or round (plastic) caps from the top of each fork leg and the Allen-headed alloy inner plugs (where fitted). Add 190 cc of either Automatic Transmission Fluid (ATF) or a proprietary fork oil to each fork leg. Use one or the other and do not attempt to mix them. Refit the inner plugs (where fitted) using jointing compound. Refit and tighten the hexagonal or round caps. Finally, refit the handlebars.

6 Steering head bearing check

Place the machine on the centre stand so that the front wheel is clear of the ground. If necessary, place blocks below the crankcase to prevent the motorcycle from tipping forwards.

Grasp the front fork legs near the wheel spindle and push and pull firmly in a fore and aft direction. If play is evident between the upper and lower steering yokes and the head lug casting, the steering head bearings are in need of adjustment. It is a good idea to ask a friend to place the fingers of one hand lightly around the steering head lug and the dust cover of the top bearing whilst the fork legs are being moved; any play that is evident will be felt by the fingers and is quite unmistakable.

To adjust the bearings, place the machine on the centre slacken the pinck bolt at the rear of the fork top yoke and tighten the large sleeve nut on the fork stem until any play is just taken up.

Take great care not to overtighten the nut. It is possible to place a pressure of several tons on the head bearings by over-

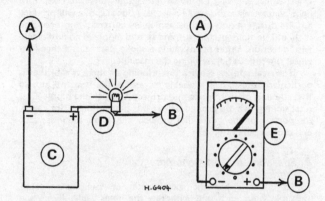

Fig. 9.7 Static timing check using battery and bulb or multimeter

A Connect to moving contact terminal
B Connect to earth (ground)
C Torch battery
D Bulb
E Multimeter set on resistance scale

Front fork leg filling point, inner plug

tightening, even though the handlebars may seem to turn quite freely. Overtight bearings will cause the machine to roll at low speeds and give imprecise steeting. Adjustment is correct if there is no play in the bearings and the handlebars swing freely and smoothly from lock to lock. Only a light tap on each handlebar end should cause them to swing. Should any roughness or hesitation be felt in the movement, then one or both of the bearings should be suspected of being damaged and should therefore be removed for further examination. Remember to retighten the fork yoke pinch bolt to its specified torque loading on completion of adjustment.

7 Wheel spoke tension check

On machines fitted with conventional steel-spoked wheels it is recommended that a frequent examination of each wheel is carried out at regular service intervals to check for loose or broken spokes. Where several spokes are broken or bent then it is recommended that the affected wheel be removed from the machine and placed in the hands of a professional wheel builder.

The easiest method of checking spoke tension is to spin the wheel whilst allowing the tip of a long screwdriver to come into light contact with the spoke centres. A loose spoke will produce a quite different note to those which are correctly tensioned and it should be tightened by turning its end nipple in an anti-clock-wise direction. Upon retensioning a spoke, carry out a check for wheel rim runout before riding the machine.

If several spokes require retensioning or there is one that is particularly loose, it is advisable to remove the tyre and tube so that the end of each spoke that projects through the nipple after retensioning can be ground off. If this precaution is not taken, the portion of the spoke that projects may chafe the inner tube and cause a puncture.

8 Speedometer and tachometer cable lubrication

To grease either the speedometer or tachometer cable, uncouple both ends and withdraw the inner cable. (On some model types this may not be possible in which case a badly seized cable will have to be renewed as a complete assembly). After removing any old grease, clean the inner cable with a petrol soaked rag and examine the cable for broken strands or other damage. Do not check the cable for broken strands by passing it through the fingers or palm of the hand, this may well cause a painful injury if a broken strand snags the skin. It is best to wrap a piece of rag around the cable and pull the cable through it, any broken strands will snag on the rag.

Regrease the cable with high melting point grease, taking care not to grease the last six inches closest to the instrument head. If this precaution is not observed, grease will work into the instrument and immobilise the sensitive movement.

Annually or every 12000 miles (19300 km)

1 Steering head bearing lubrication

The steering head bearings are packed with grease during assembly and will require repacking after the machine has reached this stage in the Routine Maintenance schedule. A considerable amount of dismantling is necessary in this instance; refer to the instructions given in Chapter 6, Sections 2 and 4.

It should be noted that the steering head bearings are now of the taper roller type with the rollers caged. Always use grease of the lithium-base, high melting point type for lubrication.

2 Wheel bearing lubrication

To remove the wheel bearings for greasing, a certain amount of dismantling is necessary. Refer to Section 7 of Chapter 7 for information about the front wheel bearings and Section 10 of the same Chapter in the case of the rear wheel bearings. Note that the threaded bearing retainers may have either a right or a left-hand thread, the latter usually being identified by the initials LH stamped on the outer face. If the retainer is driven in the wrong direction, a cracked hub may result. Always use a grease of the lithium-base, high melting point type for lubrication.

After removing the bearing retainers, check to see if the wheel is fitted with sealed bearings. These bearings, which do not require external lubrication, are fitted to later models and can be recognised by the plastic covers fitted on each side of the ball races. It may be considered advantageous to replace the earlier type of bearings with sealed bearings to obviate further lubrication periods.

3 Headlamp bulb renewal

It should be noted that Triumph recommend renewal of the headlamp bulb at frequent service intervals. Whereas it goes against the grain to reject a component which appears perfectly serviceable, this recommendation should be taken seriously, especially if the machine is being used regularly at night. At the very least, carry a spare bulb along with the tools necessary to effect bulb renewal.

4 Engine unit - dismantling, examination and reassembly

1 Although the same basic dismantling, examination and re-assembly procedure should be employed as detailed in Chapter 1, it should be noted that the model types covered by this Chapter differ in the following respects.
2 A frame that contains an integral oil reservoir is used, commonly referred to as the oil-in-frame type. In consequence, the older, side-mounted oil tank is no longer required. The base of the large diameter frame tube contains a cover retained by four nuts and washers that houses the detachable gauze filter and drain plug.
3 The primary chain is of the triplex row type. Although the chaincase needs to have its oil content added initially, it is kept topped up by the engine breather system.
4 The cylinder head is retained by ten bolts, two of which are in fact, Allen screws. Refer to the accompanying illustration for the recommended slackening and tightening sequence which, if observed, will greatly reduce the risk of the head becoming distorted.
5 The redesigned rocker boxes now have different covers. Each box has a single finned cover which is retained in position by six screws; this cover replaces the two separate threaded covers which showed a marked tendency to work loose.
6 It should be noted that the correct valve clearances are as follows:

TR7T model
Inlet 0.15 mm (0.006 in)
Exhaust 0.15 mm (0.006 in)
All other models
Inlet 0.20 mm (0.008 in)
Exhaust 0.15 mm (0.006 in)

These clearances should be set by following the special procedure described in the three monthly (3000 mile) Routine Maintenance schedule, and with the engine COLD.
7 The different type of frame used necessitates different engine plates and also a revised form of cylinder head steady. This also applies to the fuel tank mounting method, especially as a small metal strip that bridges the front underside of the tank has to be removed before the tank can be lifted from the frame. It is imperative that this strip is replaced on reassembly, as it serves a very useful function. It prevents the two halves of the tank from flexing, which would otherwise split or crack and give rise to petrol leakage that could result in a fire.

Late T140E models on

8 Models dating from around late 1978 have a redesigned cylinder head fitted. With the fitting of the new, slimmer Amal Mk II and Bing CV carburettors to twin carburettor models, the inlet ports of the head have been changed from the old 'splayed' design to a new parallel tract configuration. This has resulted in a cleaner gas flow across the combustion chambers, thus giving an improvement in fuel consumption and helping the engine design to meet those emission regulations now applicable in the US. Points to note when servicing this later type of cylinder head are as follows:
9 Upon removal of the valve guides, it will be seen that each guide is now seated by means of a circlip instead of it being shouldered. It is recommended that these clips be renewed at the same time as the guides.
10 Instead of being solidly mounted to the cylinder head, Amal Mk II and Bing CV carburettors are attached by means of a rubber hose and clips. Each hose must be free of splits and deterioration and clamped securely in position so that no air is allowed to pass through it and weaken the mixture entering the combustion chamber.
11 Before disconnecting the crankcase and oil tank breather pipes, note their fitted positions. The crankcase breather now

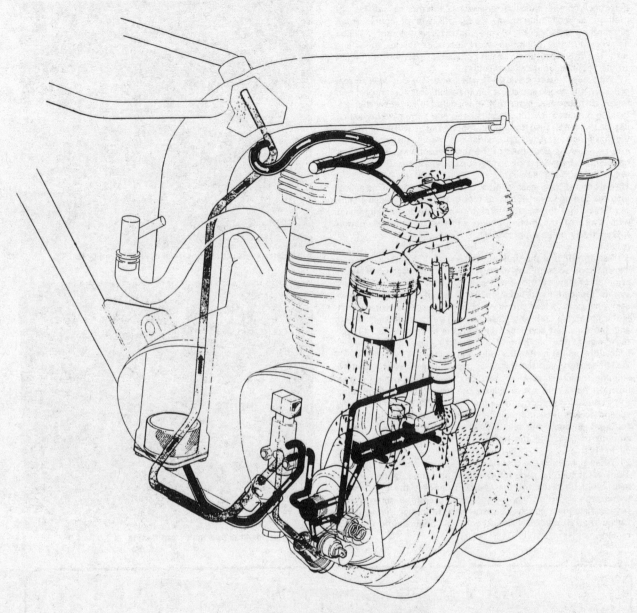

Fig. 9.8 The oil-in-frame lubrication system

feeds into the air box and the oil tank breather into the front rocker box. Both of these pipes should be free of damage or deterioration and must be correctly routed.

12 Finally, it should be noted that the exhaust pipe retaining stubs are now separate component parts which are threaded into the exhaust port outlets.

All models

13 Reference to the figure accompanying this text will show the valve timing marks for the model types covered in this Chapter. Note that these marks must be aligned exactly as shown when refitting the timing pinions.

Electric start models

14 The procedure for removal and refitting of the timing cover fitted to the engine unit of these models differs from that given in the main text of this Manual. With the battery isolated from the electrical system, disconnect the spindle terminal connection

of the oil pressure switch. Using the information given in the relevant Sections of this Chapter, remove the electronic ignition pick-up assembly and reluctor and then detach the starter motor from the timing case.

15 Work around the cover, removing each of its securing screws whilst noting their fitted position. It is a good idea to make up a cardboard template of the cover into which the screws can be inserted. Note that one of these screws will be hidden behind the starter gear pinion and this pinion should be rotated until the hole in its wall is aligned with the head of the screw, thereby allowing the screw to be removed. Ease the cover away from the engine unit whilst carefully freeing the leads of the ignition pick-up assembly from their crankcase location.

16 With the timing cover removed, take the opportunity to carry out an examination of the starter drive and sprag clutch components. Any wear in these components should be obvious but points to look out for include chipped or broken pinion teeth, excessive wear between each pinion centre and its support shaft and any loose securing nuts or cracked support plates. Note that

defective pinions must be renewed as there is no satisfactory method of reclaiming them. Doing this will of course entail dismantling part or all of the starter drive assembly. Whilst carrying out this operation, work in complete cleanliness and note the fitted position of each component part with reference to the figure accompanying this text.

17 With the starter drive dismantled and laid out in a logical order on the work surface, examine each component part and renew as necessary. Note that it is advisable to renew the self-locking nuts as a matter of course and any spring washer which has become flattened should be renewed as it will have lost its locking function. Any circlip which has become distorted should also be renewed. Each bush of the shock absorber assembly must be examined for signs of damage or deterioration. If necessary, renew the bushes as a pair.

18 Before refitting each pinion of the starter drive assembly into the timing cover, lubricate it with the recommended engine oil. Check that the teeth of each pinion are in correct alignment with those of the mating pinion and ensure that each pinion rotates freely on its support shaft.

19 Before refitting the timing cover, examine the crankshaft and camshaft oil seals in the cover for signs of damage or deterioration and renew each one as necessary. Note that the crankshaft seal is retained in its location by a circlip. The figure which accompanies this text shows which way the seals are fitted; that is the lip of the crankshaft seal faces away from the engine unit and the camshaft seal faces the opposite direction. Refitting of the cover to the engine unit is a direct reversal of the removal procedure, whilst noting the following points.

20 If necessary, replace the cover gasket with a new item before aligning the cover over its locating dowels and pushing it into position. Take care not to damage the camshaft seal whilst doing this; Triumph supply a service tool (No 61-7013) which is tapered and screws into the camshaft, thus making insertion of the camshaft through the seal a great deal easier. Apply a liberal amount of grease to the lip of each seal to help avoid damage during fitting.

21 When tightening the cover securing screws, work in a diagonal sequence tightening each screw a little at a time; doing this will help avoid distortion of the cover. Reconnect the oil pressure switch, the electronic ignition components and the starter motor before reconnecting the battery and retiming the ignition. Check that all electrical connections are clean before they are remade.

4.4 Allen screws act as additional cylinder head bolts

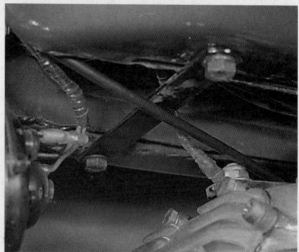

4.7 Strap prevents fuel tank from flexing

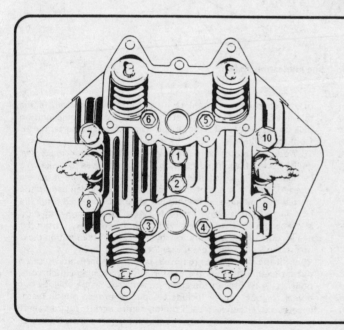

Fig. 9.9 Cylinder head bolt loosening and tightening sequence

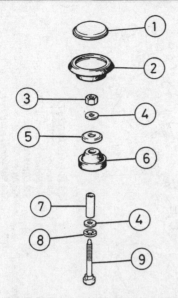

Fig. 9.10 Fuel tank mounting assembly

1 *Name plate*
2 *Plug*
3 *Nut*
4 *Washer*
5 *Spacer*
6 *Mounting rubber*
7 *Spacer*
8 *Special washer*
9 *Bolt*

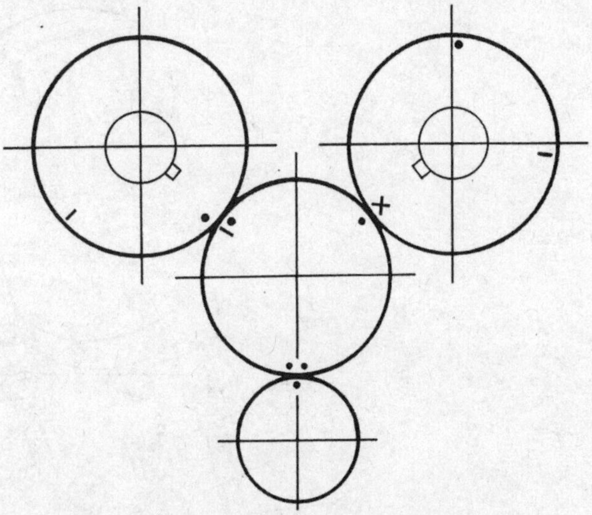

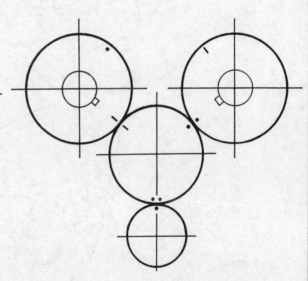

Fig. 9.11 Valve timing marks for TR65

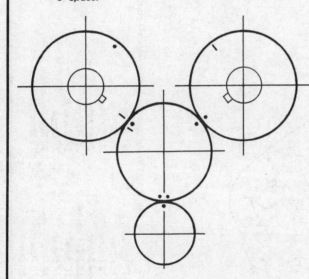

Fig. 9.12 Valve timing marks for TR65T and TR7T

Fig. 9.13 Valve timing marks for T140 and TR7

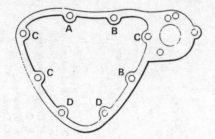

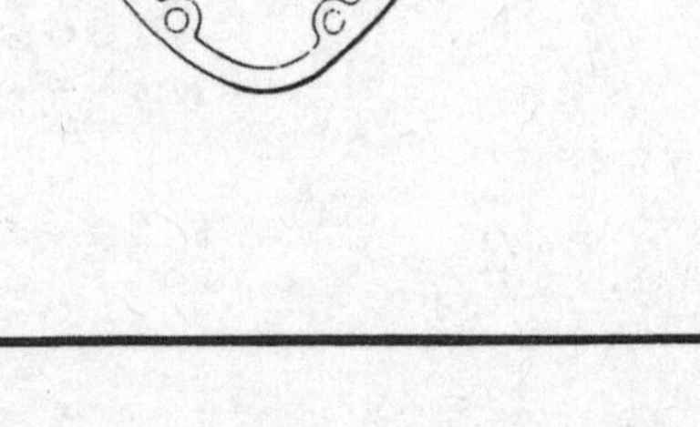

Fig. 9.14 Timing cover screw positions

A *2.5 in length*
B *2.75 in length*
C *1.0 in length*
D *1.25 in length*

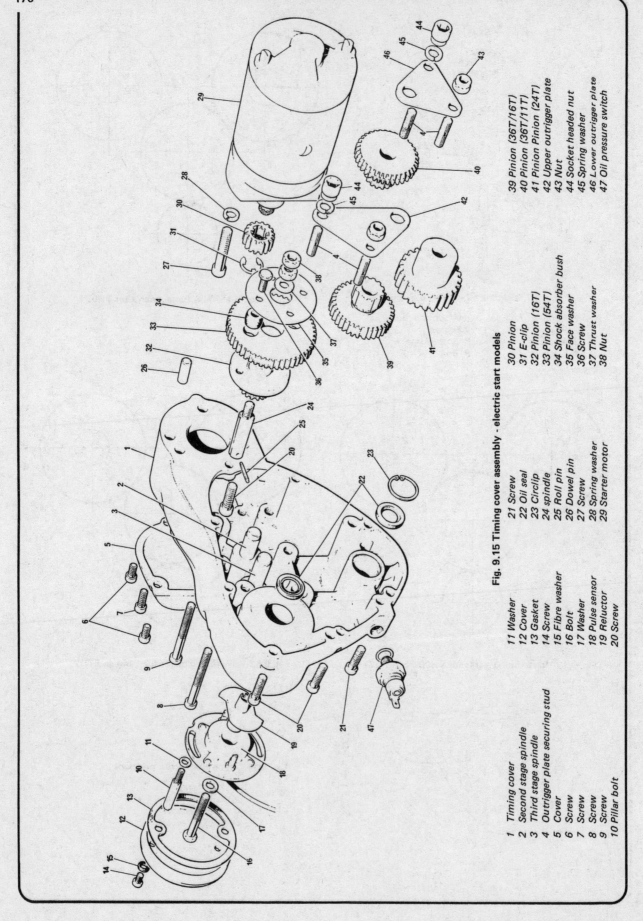

Fig. 9.15 Timing cover assembly - electric start models

1 Timing cover	11 Washer
2 Second stage spindle	12 Cover
3 Third stage spindle	13 Gasket
4 Outrigger plate securing stud	14 Screw
5 Cover	15 Fibre washer
6 Screw	16 Bolt
7 Screw	17 Washer
8 Screw	18 Pulse sensor
9 Screw	19 Reluctor
10 Pillar bolt	20 Screw

21 Screw	30 Pinion
22 Oil seal	31 E-clip
23 Circlip	32 Pinion (16T)
24 spindle	33 Pinion (54T)
25 Roll pin	34 Shock absorber bush
26 Dowel pin	35 Face washer
27 Screw	36 Screw
28 Spring washer	37 Thrust washer
29 Starter motor	38 Nut

39 Pinion (36T/16T)
40 Pinion (36T/11T)
41 Pinion Pinion (24T)
42 Upper outrigger plate
43 Nut
44 Socket headed nut
45 Spring washer
46 Lower outrigger plate
47 Oil pressure switch

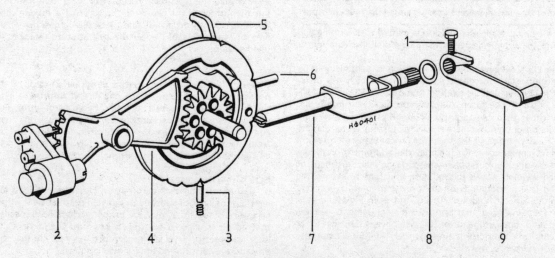

Fig. 9.16 Left-hand gearchange selector components

1	Pinch bolt	4	Plunger quadrant	7	Footchange spindle
2	Quadrant	5	Selector fork	8	O-ring
3	Plunger	6	Selector spindle	9	Pedal

5 Gearbox - dismantling, examination and reassembly

1 The gearbox is virtually identical to that fitted to the earlier models, apart from the fact that a cross-over gearchange is now employed. Instead of taking the gearchange lever shaft out of the right-hand inner and outer covers, it is now extended backwards through the inner cover and arranged to run through the inner and outer primary chaincase covers, via a simple but ingenious cranked link around the clutch chainwheel, see the accompanying figure. The positive stop and gear selector mechanism remains in its original position.

2 This changeover has necessitated transferring the rear brake pedal to the right-hand side of the machine, where it is linked directly to the rear disc brake master cylinder.

3 There is a coupling in the gearchange lever shaft, to simplify removal of either the gearbox or the inner half of the primary chaincase.

6 Clutch - dismantling, examination and reassembly

1 As mentioned in the preceding Section, a coupling in the gearchange lever shaft aids the removal of the cranked link around the clutch chainwheel, to facilitate removal of the clutch itself. A triplex row chain is now used for the primary transmission, and in consequence it is not possible to utilise components from any of the earlier models.

2 When attempting to gain access to the rubbers of the clutch shock absorber assembly, it is first necessary to determine by which method the cover plates are attached to the clutch centre. The figure which accompanies this text clearly shows the method of cover plate attachment used after that shown in Chapter 3. Note that after 1978 the three bolts securing the outer cover plate to the clutch centre were disposed with. With this change, the ends of the fixing pins were peened in such a way that they took on the appearance of screw heads and this has led to the pins being mistaken for the countersunk screws mentioned in Chapter 3. Check carefully before going ahead with cover plate removal.

3 Note that if the primary chaincase is removed for any reason, then it must be replenished with 150 cc (5.28/5.07 Imp/US fl oz) of SAE 20W/50 engine oil. Thereafter topping up is unnecessary as this is achieved through the action of the engine breather.

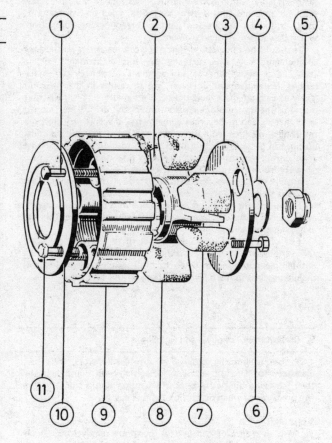

Fig. 9.17 Clutch centre assembly

1	Fixing pin		7	Rebound rubber (small)
2	Shock absorber spider		8	Drive rubber (large)
3	Outer cover plate		9	Clutch centre
4	Washer		10	Lower cover plate
5	Nut		11	Bolt
6	Bolt (pre-1978 type)			

7 Carburettor adjustment and exhaust emissions - general note

In some countries legal provision is made for describing and controlling the types and levels of toxic emissions from motor vehicles.

In the USA exhaust emission legislation is administered by the Environmental Protection Agency (EPA) which has introduced stringent regulations relating to motor vehicles. The Federal law entitled The Clean Air Act, specifically prohibits the removal (other than temporarily) or modification of any component incorporated by the vehicle manufacturer to comply with the requirements of the law. The law extends the prohibition to any tampering which includes the addition of components use of unsuitable replacement parts or maladjustment of components which allows the exhaust emissions to exceed the prescribed levels. Violations of the provisions of this law may result in penalties of up to S10 000 for each violation. It is strongly recommended that appropriate requirements are determined and understood prior to making any change to or adjustments of components in the fuel, ignition, crankcase breather or exhaust systems.

To help ensure compliance with the emission standards some manufacturers have fitted to the relevant systems fixed or pre-set adjustment screws as anti-tamper devices. In most cases this is restricted to plastic or metal limiter caps fitted to the carburettor pilot adjustment screws, which allow normal adjustment only within narrow limits. Occasionally the pilot screw may be recessed and sealed behind a small metal blanking plug, or locked in position with a thread-locking compound, which prevents normal adjustment.

It should be understood that none of the various methods of discouraging tampering actually prevents adjustment, nor, in itself, is re-adjustment an infringement of the current regulations. Maladjustment, however, which results in the emission levels exceeding those laid down, is a violation. It follows that no adjustments should be made unless the owner feels confident that he can make those adjustments in such a way that the resulting emissions comply with the limits. For all practical purposes a gas analyser will be required to monitor the exhaust gases during adjustment, together with EPA data of the permissible Hydrocarbon and CO levels. Obviously, the home mechanic is unlikely to have access to this type of equipment or the expertise required for its use, and therefore, it will be necessary to place the machine in the hands of a competent motorcycle dealer who has the equipment and skill to check the exhaust gas content.

For those owners who feel competent to carry out correctly the various adjustments, specific information relating to the anti-tamper components fitted to the machines covered in this manual is given in the relevant Sections of this Chapter.

8 Carburettors - removal and refitting

1 Before attempting removal of the carburettors, it is first necessary to gain full access by detaching both side panels and then removing the petrol tank. Remember to take the necessary fire precautions when disturbing the fuel system.

Amal Mk II

2 With this type of carburettor, continue removal by detaching both of the air filter housing covers. Completely remove the left-hand securing bolt of the cross member and then withdraw each choke plunger assembly from its carburettor housing.

3 Move to the left-hand carburettor and disconnect the fuel feed pipe from the carburettor body. Where applicable, withdraw the retaining clip from the right-hand choke linkage before disconnecting the linkage. Loosen the two jubilee clips which retain each flexible mounting hose to the cylinder head and carburettor.

4 Grasping the right-hand carburettor, ease it clear of its location then unscrew the cap from the top of the mixing chamber. Withdraw the throttle slide and place the carburettor on a clean area of work surface. Repeat this operation to remove the left-hand carburettor.

Bing CV

5 With this type of carburettor, continue removal by disconnecting the end nipples of the choke and throttle cables from their respective attachment points on each carburettor. Pull the cables clear of each carburettor and then detach the fuel feed and balance pipes from one carburettor. Free each carburettor from its flexible mounting hose by loosening the attachment clip and then ease each carburettor clear of the machine.

Both carburettor types

6 Before refitting each carburettor, closely examine its mounting hose for signs of splitting or perishing and renew it if thought necessary. Each hose must be properly clamped in position so that no air is allowed to find its way past the joints. Refitting each carburettor is a straightforward reversal of the removal procedure, whilst noting the following points.

7 Before starting the engine, check that all disturbed throttle and choke controls function smoothly over their full operating range. Check all disturbed fuel feed connections for leaks. On no account should fuel be allowed to come into contact with hot engine castings as fire may result causing personal injury. With the engine running at full operating temperature, pay attention as to whether both carburettors are correctly synchronised.

9 Amal Mk II carburettors - adjustment settings

1 Refer to the Specifications at the beginning of this Chapter for the carburettor settings.

2 To adjust the throttle stop screw, set it so that the throttle slide is opened just enough to keep the engine running at a slow tick-over with the throttle twistgrip closed.

3 The pilot air screw controls the suction on the pilot jet. It achieves this by metering the volume of air to be mixed with petrol. The idling mixture is set by turning the screw in to enrich the mixture or out to weaken it. Set the screw 2½ turns out from the fully-in position to achieve a basic setting.

4 Do not attempt to adjust the needle and needle jet setting without first obtaining advice from a Triumph specialist. These settings are made before the machine leaves the factory.

5 The amount of throttle slide cut-away is indicated by a number on the base of the throttle slide, eg. 2928. The 2½ refers to the amount of cut-away. A smaller cut-away (eg. 2) gives a richer mixture whereas a larger cut-away (eg. 4) gives a weaker mixture. The 2928 refers to the throttle type.

8.6 Check that each carburettor is properly refitted (Bing CV)

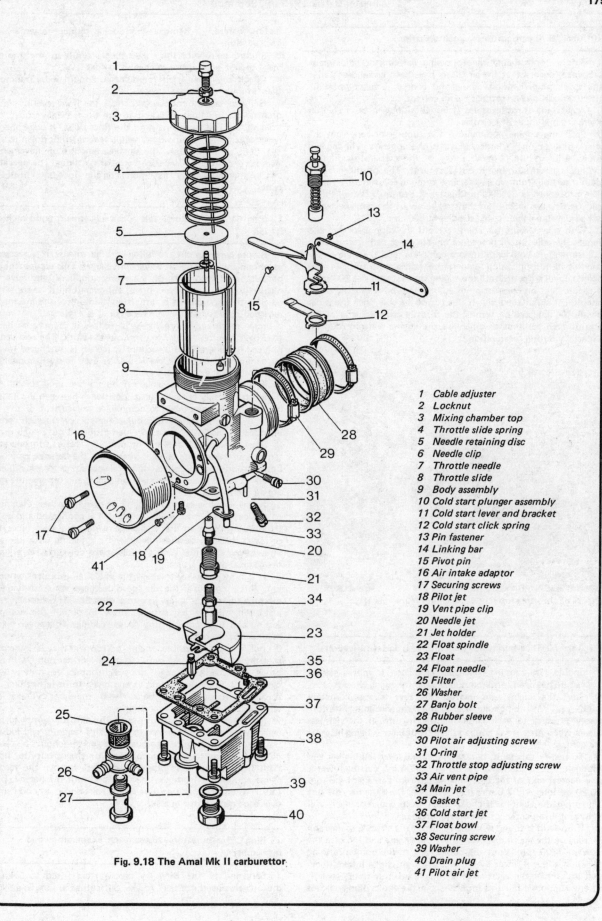

Fig. 9.18 The Amal Mk II carburettor

1 Cable adjuster
2 Locknut
3 Mixing chamber top
4 Throttle slide spring
5 Needle retaining disc
6 Needle clip
7 Throttle needle
8 Throttle slide
9 Body assembly
10 Cold start plunger assembly
11 Cold start lever and bracket
12 Cold start click spring
13 Pin fastener
14 Linking bar
15 Pivot pin
16 Air intake adaptor
17 Securing screws
18 Pilot jet
19 Vent pipe clip
20 Needle jet
21 Jet holder
22 Float spindle
23 Float
24 Float needle
25 Filter
26 Washer
27 Banjo bolt
28 Rubber sleeve
29 Clip
30 Pilot air adjusting screw
31 O-ring
32 Throttle stop adjusting screw
33 Air vent pipe
34 Main jet
35 Gasket
36 Cold start jet
37 Float bowl
38 Securing screw
39 Washer
40 Drain plug
41 Pilot air jet

10 Amal Mk II carburettors - synchronisation

1 Before commencing the following carburettor adjustment sequence, ensure that the air filters have been cleaned and that the spark plug, ignition timing and contact breaker point (if fitted) condition and settings are all correct.
   Adjust the throttle cables (from the junction box) so that there is a minimum of free play.
2 Start the engine and remove the suppressor cap from the spark plug of one cylinder. Move to the opposite cylinder and screw fully in the pilot air screw of the carburettor of that cylinder before turning it out 2½ turns. The position is the datum setting from which the slow running adjustment is made. The procedure is detailed in Chapter 4, Section 7. When this is satisfactory adjust the throttle stop screw of the same carburettor so that the engine speed steadies at 500 rpm.
3 With one carburettor adjusted, refit the suppressor cap and repeat the adjustment procedure on the other carburettor. With adjustment of both carburettors complete, and the engine running at its full operating temperature, adjust both throttle stop screws equal amounts until the engine speed steadies at 800 rpm.
4 Note that on completion of adjustment, both throttle slides should lift simultaneously. If this is not the case, then the slides should be adjusted by turning the throttle cable adjuster at the top of each carburettor otherwise the engine will run roughly, especially during acceleration.

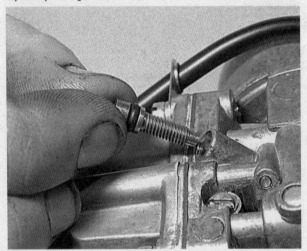

10.2 The throttle stop screw and O-ring (Amal Mk II)

11 Amal Mk II carburettors - fault diagnosis and rectification

1 Should rich running be encountered after a normal setting procedure has been carried out, make the following checks.
2 Start the engine and increase the revs to approximately 1300 rpm. With the choke lever in the normal running position, apply pressure to each choke plunger in turn. If the richness clears when pressure is applied then the plunger is being held off its seating.
3 To rectify this fault, remove the choke lever linking bar and elongate one of the holes by 1.6 mm (0.06 in), cutting towards the nearest end of the bar. This should result in a slot 4.8 mm (0.20 in) long and 3.0 mm (0.12 in) wide. Refit the bar, the slot operating the right-hand choke lever. Both choke plungers will now seat, irrespective of carburettor angles.
4 If no fault is found when carrying out the check in paragraph 2, remove the air filter covers, start the engine and increase the revs to 1300 rpm. With the choke lever in the normal running position, place a finger over the primary air choke intake (located just under the right-hand air intake adaptor fixing screw). If any suction is felt this indicates that the choke plunger is not

seating correctly. Remove the choke plunger assembly for investigation.
5 Should neither of the above checks result in any improvement then it is necessary to check the float level.
6 Check the float itself for damage. Should it be punctured then a replacement is necessary.
7 Remove each float bowl and check the float level by lightly pressing on the float, adjacent to the float needle groove. The float should rest horizontally in the float bowl. If adjustment is necessary, gently tap the float needle seating up or down to alter the float position. Only the smallest amount of movement is necessary due to the leverage ratio of the float. The operation can be made easier by first immersing the float bowl in boiling water.

12 Bing CV carburettors - idle speed adjustment and synchronisation

1 Before carrying out the following adjustments, it is necessary to obtain the use of a set of vacuum gauges. This equipment can be purchased from one of the many suppliers who advertise regularly in the motorcycle press, although it must be appreciated that this equipment is not cheap, and unless the machine is regarded as a long-term purchase, or it is envisaged that multi-cylinder motorcycles are likely to follow it, it may be better to allow a Triumph dealer to carry out the work. The cost can be reduced considerably if a vacuum gauge set is purchased jointly by a number of owners. As it will be used fairly infrequently, this is probably a sound approach.
2 Position the vacuum gauge set on the machine so that the dials can be easily read; the usual position is between the handlebars. It is now necessary to commence adjustment of the idle speed by turning the mixture adjustment screw on each carburettor fully in against its seat and then turning it out ¾ of a turn.
3 Start the engine and run it until it reaches full operating temperature. Stop the engine, disconnect the balance pipe from both carburettors and then connect a vacuum gauge pipe to each of the exposed nozzles. Restart the engine and observe the reading on each gauge.
4 If one gauge is seen to read higher than the other, then move to the carburettor to which that gauge is connected and adjust its idle stop screw so that the gauge readings equalise. Now raise or lower the idle speed as necessary by adjusting both idle stop screws even amounts. With the idle speed correctly set, equalise the gauge readings again.
5 It is now necessary to check the petrol/air mixture composition. This is correct if the idle speed decreases when turning each mixture adjustment screw in either direction. If necessary, readjust each mixture adjustment screw to obtain the optimum setting and then equalise the gauge readings. Idle speed adjustment is now complete.
6 For the best possible engine performance, it is imperative that the carburettors are working in perfect harmony with each other. If the carburettors are not synchronised, not only will one cylinder be doing less work, at any given throttle opening, but it will also in effect have to be carried by the other cylinder. This will reduce performance accordingly.
7 To check synchronisation, open the throttle gently to raise engine speed. Any difference in the gauge readings will indicate the need for throttle cable adjustment and this should be carried out by rotating the adjuster mounted on the carburettor body. With both carburettors properly adjusted, remove the vacuum gauge equipment from the machine and refit the balance pipe. Check to ensure that this pipe is not split and is a good firm fit over each carburettor nozzle.

13 Bing CV carburettors - dismantling, examination and reassembly

1 Servicing of the Bing CV carburettors fitted to Triumph motorcycles need not lead to any difficulties as long as a degree

of common sense is applied and due reference made to the diagram of the carburettor component parts which accompanies this text. It is strongly advised that each carburettor be dismantled and reassembled separately, to prevent accidentally interchanging the components. Dismantle and examine each carburettor, following an identical procedure as described below. Before any dismantling work takes place, drain out any residual fuel and clean the outside of the instruments thoroughly. It is essential that no debris finds its way inside the carburettors. Remember to observe the necessary fire precautions when draining each carburettor.

2   Remove the vacuum chamber cover retaining screws and lift the cover from position. Carefully lift the vacuum piston assembly out of the carburettor body and place it on a clean area of work surface. Detach the spring from the piston. If the diaphragm is seen to be split or perished, then it should now be removed from the piston by unscrewing the four screws with lock washers which hold its retaining ring in place. Before placing the piston to one side, inspect it for signs of wear or damage and check that it moves freely in the carburettor body. Check also for wear between the piston needle and the jet in which it slides. Wear should not occur between these two components until they have been in service for a considerable amount of time but once it does occur, petrol consumption will begin to increase. If necessary, the needle can be removed from the piston by holding it between thumb and forefinger and then gently pulling it whilst imparting a rotary motion to help ease it from position. Remember to note the fitted position of the needle by counting the number of clicks as the needle passes out of its retainer. Never interchange the piston or cover of one carburettor with that of another.

3   Invert the carburettor and unclip the float chamber. Lift the chamber from the carburettor body and note its sealing gasket. This gasket need not be disturbed unless it is damaged. The twin float assembly can be removed after displacing the pivot pin. Check the condition of the floats. If they are damaged in any way, they should be renewed. The float needle and needle seating will wear after lengthy service and should be inspected carefully. Wear usually takes the form of a ridge or groove which will cause the float needle to seat imperfectly.

4   Unscrew and remove the main jet followed by the needle jet holder. When unscrewing any jet, a close fitting screwdriver or spanner must be used to prevent damage to the soft jet material. The needle jet should fall from its housing when the carburettor is tilted towards its fitted position. If the jet proves difficult to remove by this method then it must be carefully pushed out from the carburettor venturi side by using a small soft-wood block or something that will not damage the jet face. Unscrew and remove the idling jet.

5   Before removing the mixture adjustment screw, screw it fully in until it seats, gently counting and recording the number of turns required to do so. Failure to observe this precaution will make it necessary to reset the carburettor screw settings on reassembly. Unscrew and remove the mixture adjustment screw followed by its spring.

6   It is not recommended that the throttle valve assembly be removed as the valve itself is not subject to wear. If wear has occurred on the operating pivots, a new carburettor will be required as air will find its way along the pivot bearings resulting in a weak mixture.

7   The cold start assembly is secured to the side of the carburettor body by four screws. Take care when detaching the assembly to avoid tearing its sealing gasket. Any wear between the component parts of the assembly will be obvious and any unserviceable items must be removed.

8   Clean each carburettor part thoroughly in clean petrol before placing it on a piece of clean rag or paper. Use a soft nylon brush to remove any stubborn contamination on the castings and blow dry each part with a jet of compressed air. Avoid using a piece of rag for cleaning since there is always a risk of particles of lint obstructing the airways or jet orifices. Never use a piece of wire or any pointed metal object to clear a blocked jet, it is only too easy to enlarge a jet under these circumstances and increase

the rate of petrol consumption. If an air line is not available, a blast of air from a tyre pump will usually suffice. If all else fails to clear a blocked jet, remove a bristle from the soft-bristled brush and carefully pass it through the jet to clear the blockage. Remember to observe the necessary fire precautions during the cleaning procedure and take care to guard against any blowback of petrol by wearing eye protection.

9   On completion of cleaning, check each casting for cracks or damage and check that each mating surface is flat by laying a straight-edge along its length. Any distorted casting must be replaced with a serviceable item.

10 Remove all unserviceable O-rings and sealing gaskets from the component parts and replace them with new items. Ensure that, where applicable, they are correctly seated in their retaining grooves. Any spring washers that have become flattened should now be renewed. All springs should be carefully examined for signs of corrosion and fatigue and renewed if necessary.

11 Prior to reassembly of each carburettor, check that all of its component parts, both new and old, are clean and laid out on a piece of clean rag or paper in a logical order. On no account use excessive force when reassembling the carburettor because it is easy to shear a jet or some of the smaller screws. Furthermore, the carburettor is cast in a zinc based alloy which itself does not have a high tensile strength. If any of the castings are damaged during reassembly, they will almost certainly have to be renewed.

12 Reassembly is basically a reversal of the dismantling procedure, whilst noting the following points. If in doubt as to the correct fitted position of a component part then refer to the figure accompanying this text. Ensure that the piston diaphragm locating segments are correctly aligned with the piston and also with the carburettor body.

14 Air filter elements - removal, cleaning and refitting

1   The two air filter elements fitted to each machine are formed from a sandwich of corrugated metal gauze and surgical gauze; this sandwich being bonded between two end plates which are rubber coated; this rubber coating forms an effective seal when each element is fixed in its housing.

2   To remove each element, first detach the side panels (where fitted) and then remove the single unit with lockwasher which retains each filter housing cover in position. Note the fitted position of each element before pulling it from its housing and carrying out the following examination.

3   Carefully examine the gauze of the element for signs of splitting or separation from its end plates. Any damage of this kind will necessitate renewal of the element. The same applies if the sealing area of any one end plate is found to be damaged or perished. If the element is seen to be sound, then clean the gauze by rinsing it in clean petrol or paraffin before carefully blowing it dry with a jet of compressed air. Remember to take the necessary precautions against fire whilst using petrol and take care to protect the eyes against any blow back of petrol from the air jet.

4   Relocate each serviceable filter element in its housing whilst making sure that it is fitted the correct way up, otherwise the air supply to the carburettor(s) will be shut off. Great care must be taken when positioning each element and its housing cover to ensure that no incoming air is allowed to bypass the element. If this is allowed to happen, it will allow any dirt or dust that is normally retained by the element to find its way into the carburettor and crankcase assemblies; it will also effectively weaken the petrol/air mixture.

5   Note that if the machine is being run in a particularly dusty or moist atmosphere, then it is advisable to increase the frequency of cleaning each element. Never run the engine without both elements fitted. This is because each carburettor is specially jetted to compensate for the addition of this component and the resulting weak mixture will cause overheating of the engine with the probable risk of severe engine damage.

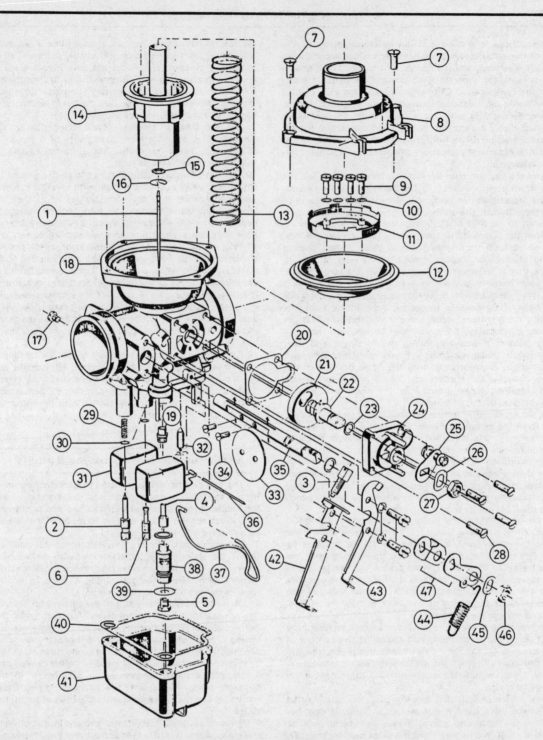

Fig. 9.19 The Bing CV carburettor

1 Jet needle
2 Mixture adjustment screw
3 Idle stop screw
4 Needle jet
5 Main jet
6 Idling jet
7 Screw
8 Vacuum chamber cover
9 Screws
10 Lock washers
11 Diaphragm retaining ring
12 Diaphragm
13 Spring
14 Piston
15 Washer
16 Clip
17 Screw
18 Carburettor body
19
20 Gasket
21 Valve disc
22 Spindle
23 Washer
24 Cold start body
25 Lever
26 Wire washer
27 Nut
28 Screws
29 Spring
30
31 Float assembly
32 Float needle
33 Throttle valve
34 Screws
35 Spindle
36 Pivot pin
37 Spring clip
38 Needle jet holder
39 Washer
30 Gasket
41 Float chamber
42 Bracket
43 Bracket
44 Spring
45 Wire washer
46 Nut
47 Lower assembly

15 Exhaust system - general

1 With the exception of the Trail models, all of the model types covered by this Chapter are fitted with a new type of long, tapered silencer which reduces the exhaust note to an acceptable level without adversely affecting engine performance to a great degree.

2 Both siamese and twin-pipe types of exhaust system are shown in the figures which accompany this text. Should a silencer require renewal, it is important that the correct replacement is obtained from a recognised Triumph spares agent. Pattern silencers will have an adverse effect on performance, as will other designs that may give the exhaust a more sporting note. Noise and speed do not necessarily go hand in hand.

3 It should also be remembered that a noisy motorcycle will have a harmful effect on the environment as a whole. On these grounds alone, tampering with the exhaust system is highly inadvisable.

16 Oil pump (4-valve) - removal and refitting

1 The fitting of this latest type of oil pump now eliminates one of the major worries of a Triumph owner, that of the machine 'sumping up' when the oil pump becomes jammed with contamination picked up by the oil as it travels round the lubrication system. Whereas if on the old 2-valve pump, one valve jamming would cause immediate trouble, on the 4-valve pump, a second valve is fitted on both the feed and return sides thus providing a fail-safe system.

2 The procedure for removal and refitting of a 4-valve pump is much the same as that given for the 2-valve pump in Section 13 of Chapter 4. When examining the pump, remember that there is a valve in the side of each plunger housing as well as at the bottom and take note of the dimension limitations and torque wrench settings given in the Specifications Section of this Chapter. The figure which accompanies this text clearly shows the fitted position of each component part of the pump assembly.

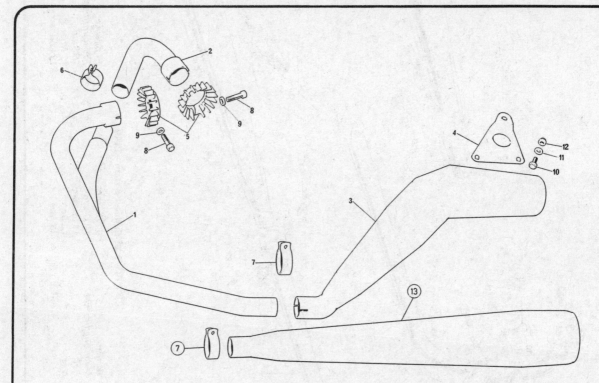

Fig. 9.20 Exhaust system - TR65 and TR7T

1 Exhaust pipe assembly	6 Clamp	10 Bolt
2 Stub pipe	7 Clamp	11 Washer
3 Silencer (TR7T)	8 Bolt	12 Nut
4 Mounting bracket	9 Washer	13 Silencer (TR65)
5 Retaining rings		

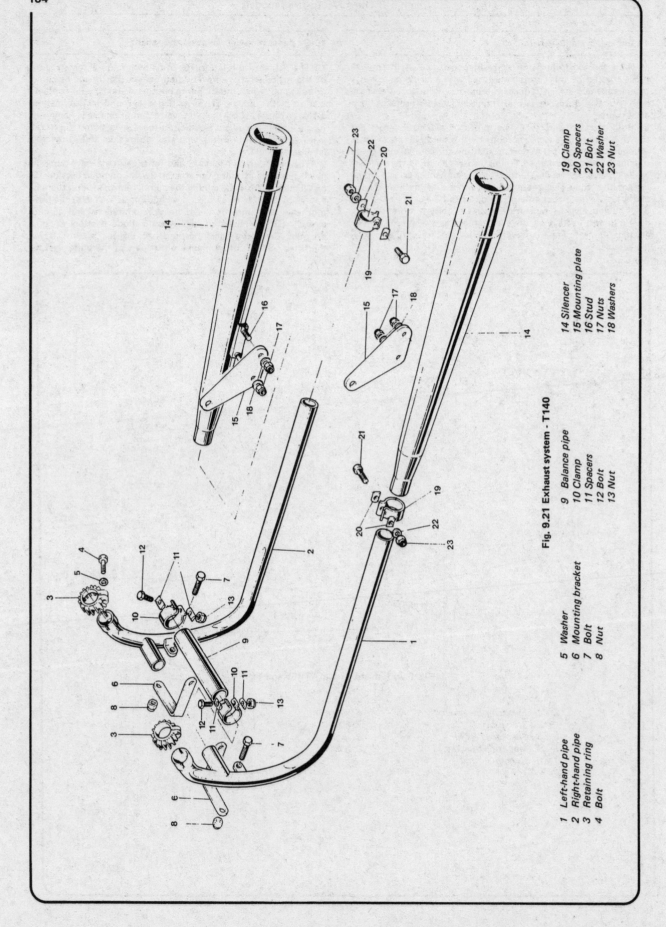

Fig. 9.21 Exhaust system - T140

1 Left-hand pipe
2 Right-hand pipe
3 Retaining ring
4 Bolt

5 Washer
6 Mounting bracket
7 Bolt
8 Nut

9 Balance pipe
10 Clamp
11 Spacers
12 Bolt
13 Nut

14 Silencer
15 Mounting plate
16 Stud
17 Nuts
18 Washers

19 Clamp
20 Spacers
21 Bolt
22 Washer
23 Nut

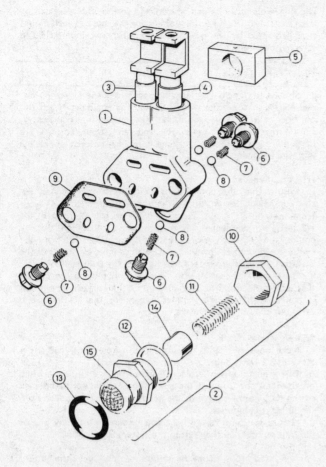

Fig. 9.22 Oil pump - 4 valve type

1 *Oil pump*	9 *Gasket*
2 *Pressure relief valve*	10 *Domed cap*
3 *Feed plunger*	11 *Spring*
4 *Return plunger*	12 *Washer*
5 *Drive block slider*	13 *O-ring*
6 *Plug - 4 off*	14 *Plunger*
7 *Spring 4 off*	15 *Filter*
8 *Ball valve - 4 off*	

17 Ignition timing - checking the setting

Contact breaker system

1 Full instructions for checking the ignition timing with the engine static are included in the Routine Maintenance Section of this Chapter. An alternative and more accurate method for checking the ignition timing can be adopted by using a stroboscopic lamp with the engine running.

2 Initial preparation will entail gaining access to the timing mark on the alternator rotor by removing the threaded inspection cap from the forward section of the primary chaincase. With this done, prepare both the rotor mark and the fixed pointer by degreasing them and then coating each one with a trace of white paint. This is not absolutely necessary but will make the position of each mark far easier to observe if the light from the 'strobe' is weak or if the timing operation is carried out in bright condition. When the light from the lamp is aimed at the timing mark and pointer, it has the effect of 'freezing' the moving mark in one position and thus the accuracy of the timing can be seen.

3 Two basic types of stroboscopic lamp are available, namely the neon and xenon tube types. Of the two, the neon type is

much cheaper and will usually suffice if used in a shaded position, its light output being rather limited. The brighter but more expensive xenon types are preferable, if funds permit, because they produce a much clearer image.

4 Connect the 'strobe' to the HT lead of the right-hand cylinder whilst following the maker's instructions. If an external 12 volt power source is required to operate the lamp, do not use the battery on the machine with it connected. This is because ac pulses in the low tension wiring of the machine can trigger the lamp thus causing it to give false readings.

5 Start the engine and increase the speed to 2500 rpm or more. Aim the strobe at the fixed pointer. Ignition timing is correct when the rotor mark is seen to be in exact alignment with the pointer; if this is not the case then switch off the engine and remove the small circular cover from the engine unit timing cover. Doing this will expose the contact breaker assembly. The two pillar bolts which retain the base plate of this assembly in position should now be loosened just enough to allow the plate to be rotated. Move the base plate a small amount, nip tight one of the pillar bolts and restart the engine. It will be seen that the position of the rotor mark has changed in relation to that of the pointer. Repeat movement of the base plate until timing mark and pointer align and then lock the plate in position by retightening the two pillar bolts.

6 Now connect the 'strobe' to the HT lead of the left-hand cylinder and repeat the timing check; this time carrying out any adjustment by moving the secondary base plate. With the check complete, refit all disturbed components and remove all test equipment from the machine.

Electronic ignition system

This type of ignition system will normally retain a perfect setting and should only need checking for correct timing if the transducer has been removed. Proceed by following the instructions given in paragraphs 2, 3 and 4 of this Section. Now set the transducer so that its fixing studs are central in their slots. With the transducer fixed in position, start the engine and increase its speed to above 3500 rpm. Note the position of the rotor mark in relation to the fixed pointer and then stop the engine. If the mark is in exact alignment with the pointer then the ignition timing is correct. Should the mark appear to the left of the pointer then the timing is retarded and should be advanced by loosening the transducer fixing studs and turning the unit in an anti-clockwise direction, or vice versa if necessary.

8 With the timing procedure complete, ensure that the transducer fixing studs are nipped tight, refit all disturbed components to the machine and remove the test equipment.

17.2 The alternator rotor timing mark and fixed pointer

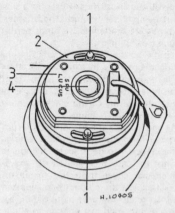

Fig. 9.23 Ignition timing unit

1 Fixing stud   3 Magnet
2 Transducer   4 Reluctor

18 Electronic ignition - general description and operation

1 The electronic ignition system fitted to Triumph motor-cycles as standard is a modified form of the Lucas Rita system. The system comprises an electronic amplifier (AB11) unit, a pick-up (5 PU) assembly, a reluctor and two ignition coils. The amplifier is a remotely mounted electronic switching system contained in a cast aluminium box. The pick-up consists of a magnetic baseplate and encapsulated winding that is mounted in the crankcase, around the reluctor. The reluctor is a precision ground steel timing device mounted on the end of the camshaft.
2 The advantages of an electronic ignition system are obvious. Because there are no mechanical contact breakers, wear does not take place and the ignition timing will remain accurate unless disturbed, or in the rare event of component failure. Although the system will retain a perfect setting, it is adjustable and full instructions for carrying out any adjustment are given in the preceding Section of this Chapter. The system operates as follows.
3 Directly the ignition is switched on, current passes through the primary windings of the two 6 volt coils which are connected in series, through the conducting amplifier and on to earth. Upon causing the crankshaft to rotate, the arms of the moving reluctor pass the poles of the pick-up and the permanent magnetic field surrounding these poles thus has its field strength rapidly changed, which in turn generates a pulse in the pick-up

winding. This pulse is then transmitted to the amplifier, causing it to switch off. High tension voltage for the spark at each plug is thus generated by the collapse of the primary winding field in each ignition coil.

19 HT coils - testing

1 Should the engine fail to start, carry out the following test procedures in the order stated. Should the engine start but run on one cylinder only, a similar test procedure is necessary. Ensure that the spark plugs, HT leads and suppressor caps are in good condition and working correctly before proceeding with the following tests.

HT output check
2 Ensure that the connections to the battery terminals are clean and tight. Switch on the ignition. Withdraw both HT leads from the coil chimneys. Remove the spark plugs and with the HT leads still connected, lay them on the engine. Hold the coil end of the HT leads approximately 6 mm (0.25 in) from the centre of the coil chimneys and turn the engine by means of the kickstart. Check for regular sparking from the coils to HT leads.
3 If no spark is seen at either coil then it must be concluded that either the amplifier, the pick-up or the coil primary circuit is faulty. Refit the spark plugs, reconnect the HT leads and carry out the following checks.

Primary circuit check
4 Make up a test lamp using a length of electrical lead and a 12 volt (2.2 watt) bulb.
5 Disconnect the white/black lead from No 2 coil and switch on the ignition. Connect the test lamp between earth and each of the four points shown in the accompanying figure. The bulb should light in all tests.
6 Should the bulb fail to light when connected to any one of the four points then the fault in the system is as follows:

Point A   No supply from the ignition switch. Check the battery supply to the switch. If the supply is correct then the switch is faulty.
Point B   No 1 coil primary open circuit. Renew the coil.
Point C   No supply to No 2 coil. Check supply lead (white/pink).
Point D   No 2 coil primary open circuit. Renew the coil.

7 Leave the test lamp connected to the No 2 coil and reconnect the white/black lead. If the lamp stays on then the fault is in the amplifier. Proceed to the next Section.

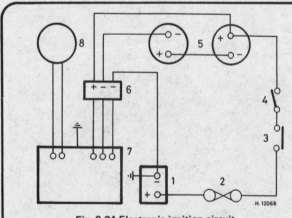

Fig. 9.24 Electronic ignition circuit

1 Battery   5 Ignition coils
2 Fuse   6 Connector
3 Ignition/lighting switch 7 Amplifier
4 Emergency switch   8 Reluctor and pick-up

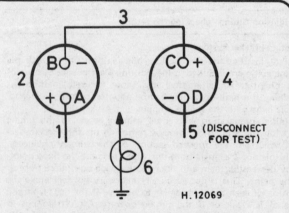

Fig. 9.25 HT coil primary circuit check

1 White/yellow lead   4 No 2 coil
2 No 1 coil   5 White/black lead
3 White/pink lead   6 Test bulb

20 Amplifier - location and testing

1 The amplifier unit for the electronic ignition system is located behind the right-hand inner side panel, below the seat. Access is gained by detaching the outer side panel, removing the air filter housing cover retaining nut and then lifting away the cover. Remove the two bolts located inside the rear edge of the air filter box and carefully withdraw the leading edge of the metal rear panel, lifting it forwards off the retaining peg. The amplifier is secured to this metal panel, a short earth lead being secured by the upper of the two securing bolts. Note also that it may be necessary to remove the air filter element to complete this operation.

2 On later models, it will be found that the amplifier has been resited on the left-hand side of the machine and is located behind the left-hand side panel, to the rear of the filter housing.

3 The only test for the amplifier is by substitution. If possible, borrow an amplifier of the same type and connect it to a good earth on the motorcycle frame. On no account operate the ignition system without earthing the amplifier case.

4 Remove the amplifier housing. Disconnect the 3-pin plug and two bullet connectors from the amplifier fitted on the machine and connect them to the substitute amplifier. Attempt to start the engine. Should the engine fail to start then there is a possible pick-up fault. Reconnect the plug and bullet connectors to the original amplifier but leave the amplifier housing removed. Proceed to the next Section.

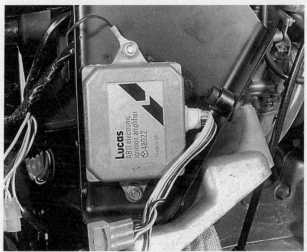

20.1 The electronic ignition amplifier unit

21 Pick-up - location and testing

1 The pick-up unit for the electronic ignition system is located behind the ribbed circular cover which is attached to the engine timing cover. To gain access to the unit, unscrew the two cover securing screws and lift away the cover and gasket. Note the fitted position of the baseplate and then remove the baseplate and windings by unscrewing the small hexagonal barrel nuts. Detach the pick-up cable bullets from the main harness and the cable from the securing straps and lift the baseplate out, carefully withdrawing the lead through the grommet in the crankcase.

2 Carry out a continuity check by disconnecting the pick-up leads at the amplifier only (white/orange and white/purple) and connecting a dc voltmeter (no load) to the battery positive terminal (red voltmeter lead) and to either of the two pick-up leads (black voltmeter lead). Earth the unconnected pick-up lead. The meter should give a reading of approximately 9 volts. Should no reading be obtained then there is an open circuit in

the pick-up winding or leads and it is necessary to proceed to the next check.

3 Carry out a pick-up wiring check by disconnecting the pick-up leads at the two bullet connectors situated under the fuel tank (ie, where the pick-up wiring is connected into the harness) and repeating the continuity check on these leads. Should no reading be obtained then the pick-up module is faulty. Should the reading be correct then the fault is probably due to a bad connection at the bullet connectors under the tank. Remake the bullet connectors ensuring that they are re-soldered if necessary and free of all corrosion and dirt.

4 Check the condition of the pick-up module by releasing the unit from the crankcase and moving it aside to allow for a substitute unit to be fitted. Temporary connections from the substitute pick-up module can be wired straight across to the amplifier. Should the engine be difficult to start or run erratically, then the ignition timing is incorrect and should be adjusted as described in Section 17.

22 Reluctor - removal and refitting

1 Remove the pick-up assembly by following the instructions given in paragraph 1 of the preceding Section. Note that if no stroboscopic lamp is readily available, then the position of the assembly base plate must be marked for reference when refitting.

2 Unscrew the centre bolt from the reluctor. The reluctor is a tapered fit on the camshaft end located by a Woodruff key and as such will require drawing from position. Triumph supply a service tool for this operation (Tool No 61-7023), but if this is not available a small two-legged puller may be used instead. If the latter method is adopted, the centre bolt should be refitted prior to the installation of the puller so that the puller centre screw bears on the bolt head and not on the easily damaged end of the camshaft. Care must be taken not to damage the reluctor in any way, because it is this component which determines the precise ignition point for each cylinder.

3 Whilst refitting the reluctor to the camshaft end, ensure that it engages correctly with the key in the taper and is pushed fully home. The specified torque loading for the centre bolt is 5 lbf ft (0.7 kgf m).

23 Frame assembly - oil-in-frame type

1 The use of the oil-in-frame type of assembly that has BSA origins has necessitated certain changes in design of some of the minor components, such as the engine mounting plates and the cylinder head steady. It has also proved necessary to modify the way in which the fuel tank is mounted.

2 Refer to the accompanying illustration for the dimensions of the new frame assembly, which will prove useful if the frame has to be checked for suspected accident damage.

3 On some machines the swinging arm pivot bolt threads into the frame. It is important to check whether this arrangement is used first, before attempting to drive the bolt out. Try turning the hexagon head, to see if the bolt will withdraw.

24 Front forks - removal, dismantling and refitting

1 Although the same basic dismantling and reassembly techniques described in Chapter 6 can be applied for removal of the front forks from any one of the model types covered in this Chapter, the task is now made somewhat easier since the forks no longer have outer shrouds and a rubber cover has replaced the original threaded dust excluder sleeve which was often difficult to remove.

2 Where a disc brake caliper is attached to a fork leg, it will of course be necessary to detach it from the triangular cast bracket

of the leg by removing its two retaining bolts with spring washers. The caliper need not be drained of fluid but great care must
be taken with this type of braking system not to operate the
brake lever once the brake disc is removed from between the
brake pads since fluid pressure may displace the piston and cause
fluid leakage. Additionally, the distance between the pads will be
reduced, thereby making refitting of the caliper over the disc
extremely difficult. To prevent any chance of this happening,
it is a good idea to place a wood wedge between the two pads
directly the caliper is removed.

3   With the hydraulic pipeline retaining bracket detached from
the fork leg, swing the caliper clear of the fork leg and tie it to
some convenient part of the machine whilst further dismantling
of the fork assembly continues. Do not risk damage to the hydraulic hose by leaving the caliper suspended by it.

4   Before refitting the caliper to the fork leg, examine the
hydraulic hose for signs of leakage, damage, deterioration or
scuffing against any of the fork component parts. Renew the
hose if it is defective. Clean and examine the caliper and its
pipeline and check the amount of wear on each of the brake
pads.

5   Finally, if in doubt as to the fitted position of any one component part of the fork assembly, then refer to the figures which
accompany this text.

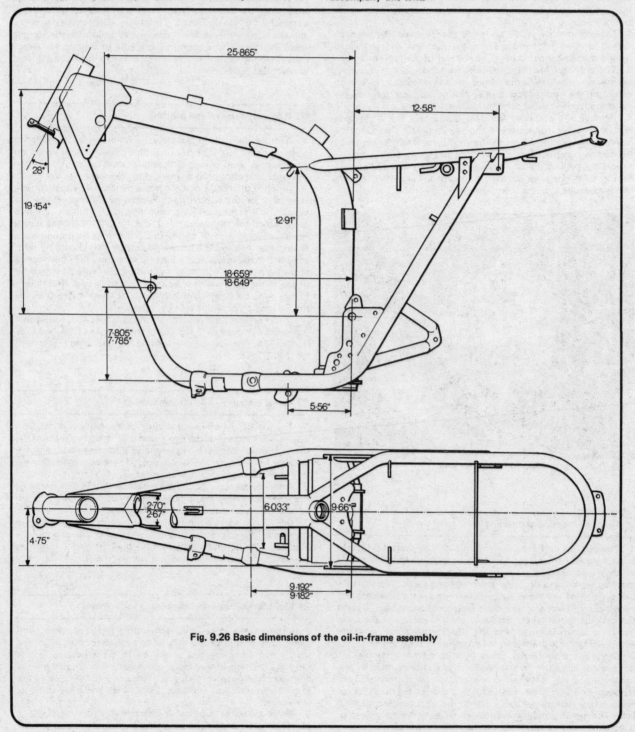

Fig. 9.26 Basic dimensions of the oil-in-frame assembly

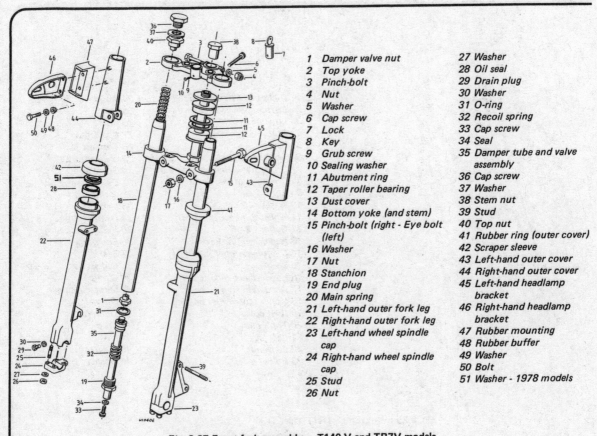

1 Damper valve nut	27 Washer
2 Top yoke	28 Oil seal
3 Pinch-bolt	29 Drain plug
4 Nut	30 Washer
5 Washer	31 O-ring
6 Cap screw	32 Recoil spring
7 Lock	33 Cap screw
8 Key	34 Seal
9 Grub screw	35 Damper tube and valve
10 Sealing washer	assembly
11 Abutment ring	36 Cap screw
12 Taper roller bearing	37 Washer
13 Dust cover	38 Stem nut
14 Bottom yoke (and stem)	39 Stud
15 Pinch-bolt (right - Eye bolt	40 Top nut
(left)	41 Rubber ring (outer cover)
16 Washer	42 Scraper sleeve
17 Nut	43 Left-hand outer cover
18 Stanchion	44 Right-hand outer cover
19 End plug	45 Left-hand headlamp
20 Main spring	bracket
21 Left-hand outer fork leg	46 Right-hand headlamp
22 Right-hand outer fork leg	bracket
23 Left-hand wheel spindle	47 Rubber mounting
cap	48 Rubber buffer
24 Right-hand wheel spindle	49 Washer
cap	50 Bolt
25 Stud	51 Washer - 1978 models
26 Nut	

Fig. 9.27 Front fork assembly — T140 V and TR7V models

1 Damper valve nut	28 Oil seal
2 Top yoke	29 Drain plug
3 Pinch-bolt	30 Washer
4 Nut	31 O-ring
5 Washer	32 Recoil spring
6 Cap screw	33 Cap screw
7 Lock	34 Seal
8 Key	35 Damper tube and valve
9 Grub screw	assembly
10 Sealing washer	36 Cap screw
11 Abutment ring	37 Washer
12 Taper roller bearing	38 Stem nut
13 Dust cover	39 Stud
14 Bottom yoke (and stem)	40 Top nut
15 Pinch-bolt (right) - Eye bolt	41 Rubber ring (outer cover)
(left)	42 Scraper sleeve
16 Washer	43 Left-hand outer cover
17 Nut	44 Right-hand outer cover
18 Stanchion	45 Left-hand headlamp bracket
19 End plug	46 Right-hand headlamp bracket
20 Main spring	47 Rubber mounting
21 Left-hand outer fork leg	48 Rubber buffer
22 Right-hand outer fork leg	49 Washer
23 Left-hand wheel spindle cap	50 Bolt
24 Right-hand wheel spindle cap	51 Oil seal retainer
25 Stud	52 Speedometer/tachometer
26 Nut	cable guide
27 Washer	

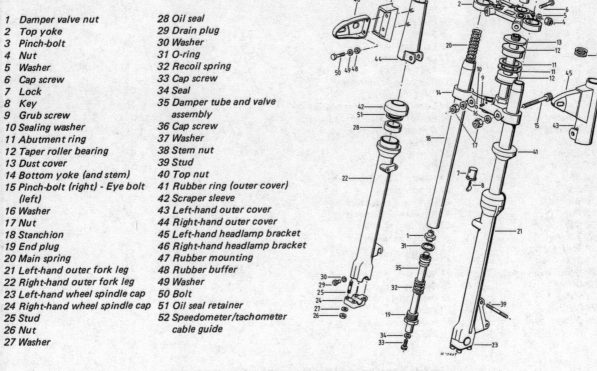

Fig. 9.28 Front fork assembly - T140E, T140D and TR7V
models

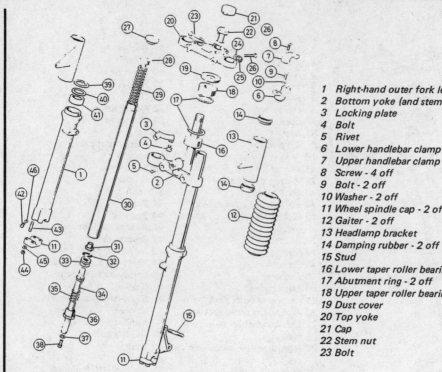

1	Right-hand outer fork leg	24	Washer
2	Bottom yoke (and stem)	25	Nut
3	Locking plate	26	Bolt
4	Bolt	27	Cap screw
5	Rivet	28	Top nut
6	Lower handlebar clamp - 2 off	29	Main spring
7	Upper handlebar clamp - 2 off	30	Stanchion
8	Screw - 4 off	31	Damper valve nut
9	Bolt - 2 off	32	O-ring
10	Washer - 2 off	33	Valve
11	Wheel spindle cap - 2 off	34	Damper tube
12	Gaiter - 2 off	35	Recoil spring
13	Headlamp bracket	36	End plug
14	Damping rubber - 2 off	37	Seal
15	Stud	38	Cap screw
16	Lower taper roller bearing	39	Circlip
17	Abutment ring - 2 off	40	Oil seal retainer
18	Upper taper roller bearing	41	Oil seal
19	Dust cover	42	Drain plug
20	Top yoke	43	Stud
21	Cap	44	Nut
22	Stem nut	45	Washer
23	Bolt	46	Washer

Fig. 9.29 Front forks - TR7T Tiger Trial

1	Right-hand outer fork leg	28	End plug
2	Top yoke	29	Seal
3	Pinch bolt	30	Cap screw
4	Nut	31	Left-hand outer cover - 1981 models
5	Washer	32	Right-hand outer cover - 1981 models
6	Bolt	33	Left-hand headlamp bracket
7	Stem nut	34	Right-hand headlamp bracket
8	Cap	35	Rubber mounting
9	Bottom yoke (and stem)	36	Rubber buffer
10	Steering lock	37	Washer
11	Key	38	Bolt
12	Spring	39	Rubber ring (outer cover)
13	Ball	40	Scraper sleeve
14	Bolt	41	Left-hand outer fork leg
15	Bolt	42	Wheel spindle clamp
16	Washer	43	Circlip
17	Nut	44	Oil seal retainer
18	Cap screw	45	Oil seal
19	Top nut	46	Stud
20	Main spring	47	Nut
21	Stanchion	48	Washer
22	Damper valve nut	49	Drain plug
23	O-ring	50	Washer
24	Valve	51	Headlamp bracket - 1982 models
25	Washer		
26	Damper tube		
27	Recoil spring		

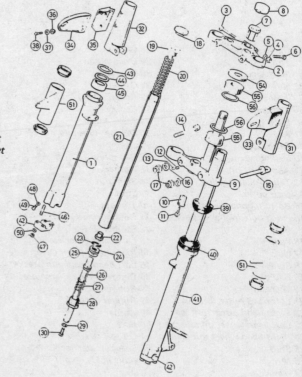

Fig. 9.30 Front forks - TR65 Thunderbird and T140 models '80 - '83

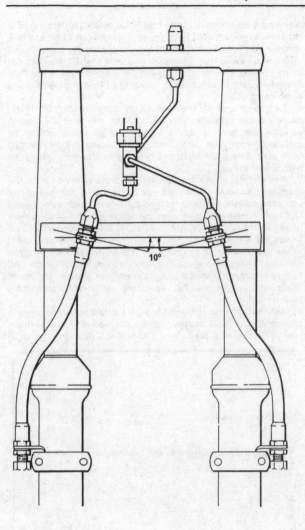

Fig. 9.31 Front brake hydraulic hose positioning

10°

25.2a Disconnect the drive cable from each instrument ...

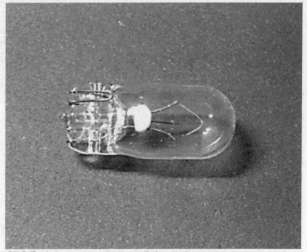

25.2b ... and withdraw each instrument from its mounting cup

25 Speedometer and tachometer heads - removal and refitting

1 With the exception of the T140V, TR7V and early T140E models, all the model types covered in this Chapter are fitted with Veglia instruments. These instruments are contained in rubber cups; the cup housings being secured to the upper fork yoke by means of a bolt passing through an attached rubber mounting, and to the base of the warning light console by two bolts and washers. The removal procedure for each instrument is identical.

2 Unscrew the drive cable to instrument retaining cap and pull the cable down and away from the instrument. The instrument can now be removed from the rubber cup by pushing up on the base of the instrument. Note that the instrument is attached to the cup by means of a small blob of glue. It is advisable to re-glue the instrument to the cup on reassembly, bearing in mind that the type of glue used must be of the type that can be re-broken; this will stop the instrument rotating within the cup.

3 With the instrument removed from the cup, pull the bulb holder and bulb out of the base of the instrument. The instrument can now be lifted away.

4 Refer to Chapter 6, Section 15, paragraph 3, and also Section 16 for details of examination and renovation. Refitting is the reverse of the removal procedure.

25.3 Each instrument bulb is of the capless type

26 Swinging arm - removal, examination and reassembly

1 If it is found necessary to remove the later type of swinging arm assembly from its frame attachments, then remove the rear wheel and proceed as follows.

2 Remove the chainguard by unscrewing its front securing bolt with washer, loosening the lower securing bolt of the left-hand rear suspension unit and then lifting the chainguard clear of its attachment points.

3 Detach each rear suspension unit from the swinging arm by withdrawing its securing bolt. Note the fitted position of the washers on the bolt. Support the swinging arm fork, remove the nut and plain washer from the fork pivot spindle and withdraw the spindle. Ease the fork rearwards to clear the machine.

4 Place the swinging arm fork on a clean work surface and dismantle the pivot assembly, pulling the spindle sleeves from their housing. Clean all the component parts in petrol or paraffin whilst observing the necessary fire precautions and lay them out in a logical order on a clean piece of rag or paper. Note the dimensions listed in the Specifications Section of this Chapter and examine the bearing surfaces of the pivot for excessive wear. Where renewal is necessary, then the bushes and/or sleeves should be renewed as a set.

5 If wear on the bushes is excessive, then they may be removed from the pivot housing by use of Triumph service tool No 61-6117. This tool and its method of use is shown in the figure accompanying this text. If the tool is not available, then a similar item can be constructed from the appropriate sized bolt and nut, washers and length of steel tubing. Alternatively, a shouldered drift can be made from mild steel and used to drift one bush out of its housing, thereby drifting out the second bush at the same time; take care to support the swinging arm close to the housing when doing this.

6 Each bush is of the steel-backed pre-sized type and therefore, when pressed carefully into its housing, will give the correct working clearance with the sleeve. Lightly smear the outer surface of each bush with grease to aid fitting and press in one bush at a time from each end of the pivot housing, using the special tool shown.

7 Reassembly of the fork pivot is a direct reversal of the dismantling procedure. If in doubt as to the fitted position of any one component part, then refer to the figure which accompanies this text. Smear each part with the recommended grease before fitting it in position. Do not omit to fit the rubber dust excluders before lifting the swinging arm fork back into the frame. These excluders must of course be free of any damage or deterioration. Tighten the spindle retaining nut to the recommended torque loading and check that the swinging arm can be moved freely throughout its full operating arc before proceeding further.

8 Lubricate the pivot assembly with a grease gun until grease is seen to appear from the pivot ends. The assembly should therefore be lubricated at least every 1000 miles (1600 km).

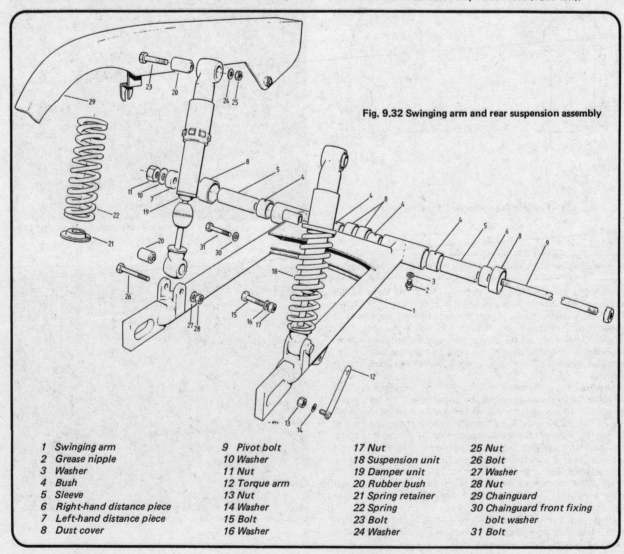

Fig. 9.32 Swinging arm and rear suspension assembly

1 Swinging arm	9 Pivot bolt	17 Nut	25 Nut
2 Grease nipple	10 Washer	18 Suspension unit	26 Bolt
3 Washer	11 Nut	19 Damper unit	27 Washer
4 Bush	12 Torque arm	20 Rubber bush	28 Nut
5 Sleeve	13 Nut	21 Spring retainer	29 Chainguard
6 Right-hand distance piece	14 Washer	22 Spring	30 Chainguard front fixing
7 Left-hand distance piece	15 Bolt	23 Bolt	bolt washer
8 Dust cover	16 Washer	24 Washer	31 Bolt

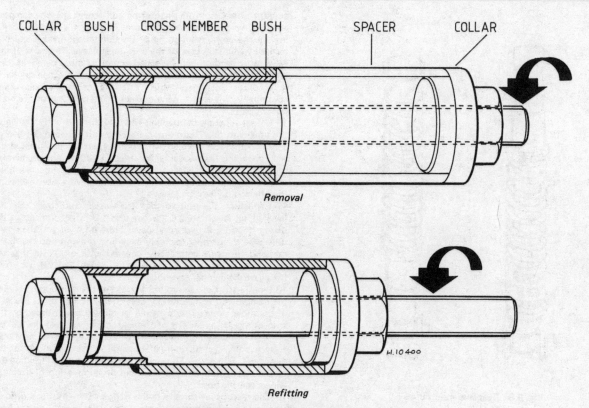

COLLAR   BUSH   CROSS MEMBER   BUSH   SPACER   COLLAR

Removal

Refitting

H.10400

Fig. 9.33 Removing and refitting swinging arm bushes

27 Rear suspension units

Girling

1   Girling gas filled suspension units are fitted to machines with frame numbers after BX05107. On no account must this type of suspension unit be mixed with the earlier oil hydraulic type. To identify the type of unit, refer to the accompanying figure whilst noting that gas filled units are equipped with an 11 mm (0.44 in) diameter damper rod whereas oil units have a 9.5 mm (0.38 in) diameter rod.

2   It is essential that each type of unit is fitted the correct way up. The gas filled unit must be fitted with the castellated adjustment ring at the top of the unit, whilst the oil hydraulic unit must be fitted with the adjustment ring at the bottom of the unit. **Note:** irreparable damage will be caused to the damping mechanism of the unit should it be fitted the wrong way up.

3   When adjusting the rear suspension units, ensure both units are adjusted equally. A quick visual check can be made for comparison from the rear of the machine.

4   Using the correct size of C-spanner, increase or decrease the spring rating as shown in the accompanying figure. For solo riding the suspension should be set at the lowest position. When carrying luggage it is recommended that the setting is in a mid-position whilst, when carrying a pillion passenger, the suspension should be set at the highest position.

5   The dismantling, examination and reassembly procedure for the suspension units is as stated in Chapter 6, Section 9 of this manual. Spring free lengths are contained in the Specifications at the beginning of this Chapter.

6   Note also, when examining the unit after dismantling, that the damper unit should be checked for the correct damping action by extending and compressing the unit by hand. If movement of the plunger rod is found to be very easy, then the unit must be replaced.

7   Check for any bending of the plunger rod. **Note:** under no circumstances should the plunger rod be lubricated. Should it be found that the bonded pivot bushes are worn or that the sleeve is loose in the rubber bush, then the bushes can be renewed by driving out the worn item and pressing in a new one, using soapy water smeared around the rubber to assist assembly.

Marzocchi

8   Two types of Marzocchi suspension are fitted as standard equipment to Triumph motorcycles; the standard unit and the Strada air assisted unit which is easily recognisable by its remote air reservoir. To service and adjust the standard unit, carry out a procedure which is similar to that given for the Girling units whilst noting the information given in the Specifications Section of this Chapter. Refer to the accompanying figure for unit adjustment and note that each unit must be fitted with its adjustment ring at the bottom of the unit.

9   The Strada unit is of the type which can be fully dismantled for servicing. If a unit is damaged or fails to operate effectively, then proceed as follows. Position the unit upright so that its lower bush housing is clamped securely between the jaws of a vice. Unscrew the cap from the air charging valve and whilst taking care to protect the eyes, release the air in the reservoir by depressing the valve insert. If oil is seen to emit from the valve, then the rubber bellow within the reservoir is defective and must be renewed. If the bellow requires renewal and the rest of the unit is fully serviceable, then proceed directly to paragraph 12 of this Section. The correct sequence for dismantling the unit as a whole is as follows.

10 With all the air released from the reservoir, slowly unscrew the oil drain screw in the side of the reservoir body until the oil stops leaking under pressure. Do not remove the screw but retighten it, finger tight. Now grasp the top of the spring with one hand and pull downwards to release the spring from its retaining collar. Detach the collar from the unit and lift the spring clear.

11 Select a spanner which is a good fit over the damper rod retaining plug, unscrew the plug and slide it clear of the unit

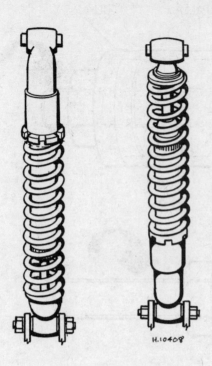

Fig. 9.34 Rear suspension units

body. Grasp the top of the damper rod and pull it clear of the unit body. Moving the rod from side-to-side whilst pulling will help to free it but if this fails, then try tapping lightly round the top section of the unit body with a soft-faced hammer. The oil drain screw can now be removed and all oil drained from the unit.

12 To remove the rubber bellow from the reservoir, unscrew the bellow retaining plug and then displace the bellow retaining cup to release the bellow. A special two-pin spanner will be required to unscrew the retaining plug. If this spanner cannot be obtained from a Marzocchi service agent, then it will have to be made up. Both the bellow and its retaining cup should be displaced by carefully using a small flat head screwdriver to lever them from position.

13 With the bellow removed, carefully examine it for damage or deterioration and renew it if necessary. The retaining plug and cup should not normally suffer damage, although the cup should be examined for cracks and the O-ring which is sited between the cup and plug automatically renewed. If the air charging valve is to be removed, then renew the O-ring. Make sure the bellow and retaining cup are both properly seated in the reservoir before fitting and tightening the retaining plug.

14 If the suspension unit has been completely dismantled for examination and servicing, then thoroughly clean the body of the unit in petrol whilst observing the necessary fire precautions. Once dry, the unit body should be carefully examined for cracks or damage which may lead to leakage of air or oil or which will restrict the passage of the damper piston in its bore. Before placing the unit body to one side ready for reassembly, ensure that its internal surfaces are clean of all contamination.

15 A bent or corroded damper rod must be renewed. Removal of the damper assembly from the rod and examination of the component parts is easily accomplished. Prepare an area of the work surface by covering it with clean rag or paper, remove the end nut from the damper rod and lay each part of the damper assembly out on the work surface in the order of removal. Renew all rubber seals as a matter of course; take note of the

amount of wear on the piston ring and renew any one component part which is damaged or worn. Carry out a similar procedure on the component parts of the compression valve which is located within the base of the unit inner sleeve. The parts of this valve must be thoroughly cleaned in petrol before being refitted. When reassembling both the valve and damper mechanism, work in absolute cleanliness whilst referring to the figures which accompany this text for the correct fitted position of the component parts.

16 Before refitting the damper rod into the unit body, replenish the unit with oil. To do this, clamp the unit between the jaws of a vice so that it is inclined 20° from the vertical, as shown in the accompanying figure. Replenish the unit with the recommended type of oil until it is seen to emit from the drain screw hole. Check that the O-ring fitted to the drain screw is serviceable and then fit and tighten the screw.

17 Reposition the unit so that it is vertical and check that the level of oil is within 2.0-2.5 cm (0.8-1.0 in) of the top of the piston housing. If necessary, correct the oil level and then carefully ease the damper assembly back into the unit body until the rebound spring is completely covered with oil. Note that the manufacturer supplies a special tool which, if used, will help fitting of the damper assembly.

18 Wrap a strip of rag around the top of the unit body. Holding the damper rod still, carefully ease the pilot boss up to the level of the inner sleeve. Any overspill of oil will be absorbed by the strip of rag. Using the flat of a small screwdriver, seat the locking seal before removing the unit from the vice and inverting it over a clean container to drain off any oil remaining on top of the pilot boss. Remember to prevent the pilot boss from sliding out of the unit whilst it is inverted and on completion of draining, wipe clean the boss.

19 The next point to note is the fitting of the dust seal onto its seat in the damper rod retaining plug. If this seal has not been renewed and shows signs of swelling through the centre of the plug with the plug tightened, then it must be renewed. Before proceeding further, check the damper rod for smooth, uninterrupted movement throughout its full operating range. If this is not the case, then remove the retaining plug and investigate the matter.

20 When refitting the spring, check that it is fitted the correct way up with its close coil end facing the bottom of the unit and that it is properly retained in position with its collar. Recharge the unit with air whilst noting the following information.

21 It must be noted that the bellow contains a very small amount of air and for this reason it is not advisable to use a compressed air supply of the type found on garage forecourts because there is a risk of damaging the bellow through overcharging. One of the various types of mini syringe pumps now available at motorcycle dealers is ideal for this purpose. These pumps are as little as five inches long and can therefore easily be carried on the machine; they come supplied with flexible connectors. Note that the recommended air change pressure for this type of unit fitted to Triumph motorcycles is 28 psi (2 kg/cm^2) whereas the maximum pressure the unit should be asked to withstand is 42 psi 3 kg/cm2.

22 It should be noted that each time the pressure gauge is removed from the valve, a small volume of air will be lost from the fork legs resulting in an equally small drop in pressure. This should be allowed for when noting the gauge reading. Before refitting the cap over the air charging valve, carry out a careful check of the complete unit for signs of air or oil leakage.

Paoli and S and W

23 At the time of writing this Chapter both of these makes of suspension unit were being fitted to Triumph TSX models as standard equipment, although S and W units were only being fitted in the US by the importers. Because the Paoli units are a recent item of equipment, no technical information is yet available unlike the S and W units which are sold in some numbers both in the UK and the US and which therefore have major distributors who can advise on the availability of spare parts and technical information.

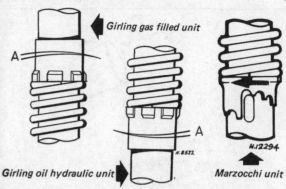

Fig. 9.35 Adjusting the suspension units

'A' denotes direction of rotation to increase spring rating

27.8 The Marzocchi Strada unit has a remote reservoir

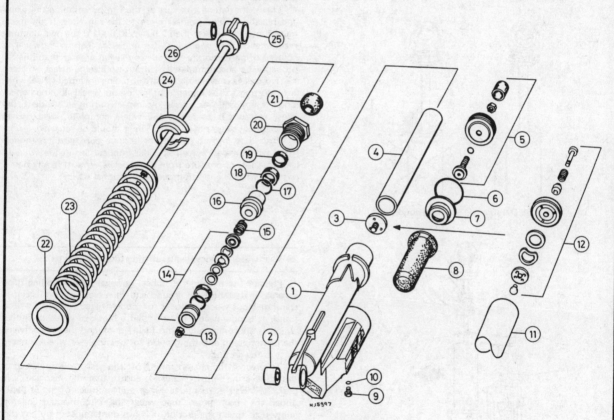

Fig. 9.36 Marzocchi rear suspension unit

1 Body and reservoir	8 Rubber bellow	15 Rebound spring	22 Spring seat
2 Lower mounting bolt	9 Oil drain screw	16 Pilot boss	23 Spring
3 Compression valve	10 O-ring	17 Sealing ring	24 Retaining collar
4 Inner sleeve	11 Unit body	18 Oil seal	25 Damper rod
5 Air charging valve	12 Compression valve assembly	19 Dust seal	26 Upper mounting bush
6 O-ring	13 End nut	20 Damper rod retaining plug	
7 Retaining cup	14 Damper assembly	21 Bump stop	

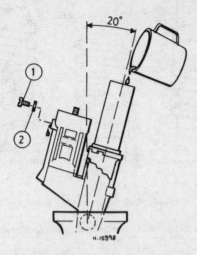

Fig. 9 37 Replenishing the Marzocchi unit with oil

1 Drain screw   2 O-ring

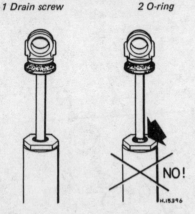

Fig. 9.38 Damper rod dust seal condition

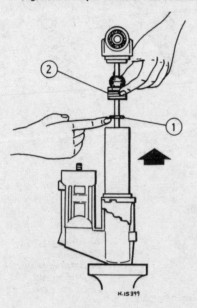

Fig. 9.39 Method of fitting the dust seal into the damper rod retaining plug

1 Dust seal   2 Plug

28 Front and rear wheels - general (cast alloy type)

1 Removal and fitting of the lester and Morris types of cast alloy wheels fitted to the model types covered by this Chapter is identical to the procedure given for the conventional wire spoked wheels within Chapter 7 of this Manual and the following Section of this Chapter.

2 Commence examination by carefully checking each wheel for cracks and chipping, particularly at the spoke roots and the edge of the rim. As a general rule a damaged wheel must be renewed as cracks will cause stress points which may lead to sudden failure under heavy load. Small nicks may be radiused carefully with a fine file and emery paper (No 600 — No 1000) to relieve the stress. If there is any doubt as to the condition of a wheel, advice should be sought from a reputable dealer or specialist repairer.

3 The surface of each wheel is protected to help prevent corrosion. If damage occurs to the wheel and the protective finish is penetrated, the bared aluminium alloy will soon start to corrode. A whitish grey oxide will form over the damaged area, which in itself is a protective coating. This deposit however, should be removed carefully as soon as possible and a new protective coating applied.

4 Check the lateral run out at the rim by spinning the wheel and placing a fixed pointer close to the rim edge. If the maximum run out is greater than 2.0 mm (0.080 in) the manufacturer recommends that the wheel be renewed. This is, however, a counsel of perfection; a run out somewhat greater than this can probably be accommodated without noticeable effect on steering. No means is available for straightening a warped wheel without resorting to the expense of having the wheel skimmed on all faces. If warpage was caused by impact during an accident, the safest measure is to renew the wheel complete. Worn wheel bearings may cause rim run out. These should be renewed.

5 Note that impact damage or serious corrosion on models fitted with tubeless tyres has wider implications in that it could lead to a loss of pressure from the tubeless tyres. If in any doubt as to the wheel's condition, seek professional advice.

29 Rear wheel - removal and refitting (disc brake type)

1 Prepare the machine for wheel removal by positioning it on an area of flat and level ground and then placing it on its centre stand or stout wooden blocks for models fitted with only a prop stand. It is now necessary to obtain a block of wood or similar which is approximately 8 cm (3 in) thick and place it between the centre stand and the ground so that the rear wheel is raised clear of the ground.

2 To prevent the chain running off the gearbox sprocket after the connecting link is removed, ensure that the machine is in gear. It is also advisable to place a clean piece of rag or paper under the machine so that when the disconnected chain is allowed to touch the ground it is not contaminated by any dirt or grit. Remove the chain connecting link and free the chain from the rear sprocket.

3 Where necessary, remove the left-hand exhaust silencer so that the wheel spindle can be withdrawn. Hold the wheel spindle with a tommy bar passed through the hole provided in the spindle end, remove the spindle retaining nut and washer and withdraw the spindle.

4 Support the wheel whilst carefully swinging the brake caliper clear of the brake disc. Ease the wheel down and rearwards until the speedometer gearbox with cable can be detached from the wheel hub to allow the wheel to be pulled clear of the machine.

5 Note that great care must be taken not to operate the brake pedal once the disc is removed from between the brake pads since fluid pressure may displace the piston and cause fluid

leakage. Additionally, the distance between the pads will be reduced, thereby making refitting of the wheel extremely difficult. To prevent any chance of this happening, it is a good idea to place a hardwood wedge between the brake pads directly the wheel is removed.

6   Refitting the wheel is a direct reversal of the removal procedure, whilst noting the following points. Ensure that the speedometer gearbox is correctly engaged with the wheel hub before lifting the wheel into position. Take great care not to damage the brake pads when sliding the disc between them. Lightly grease the length of the wheel spindle before fitting it and ensure that the caliper support is correctly retained by the spindle. Tighten the spindle retaining nut with washer, finger-tight and then reconnect the final drive chain over the rear wheel sprocket. Note that the spring clip which retains the connecting link in position must have the link side plate fitted beneath it, be seated correctly and have its closed end facing the direction of chain travel. Finally, refer to the Routine Maintenance Section of this Chapter and check the tension of the final drive chain.

29.6 The speedometer gearbox must be correctly engaged with the wheel hub

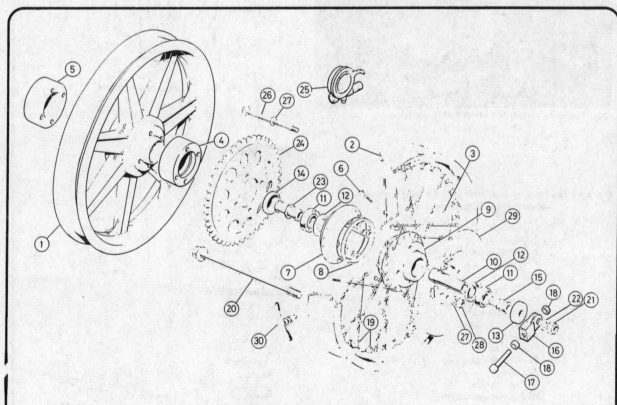

Fig. 9.40 Rear wheel - disc brake type

1 Wheel - cast alloy	9 Hub	17 Adjusting bolt - 2 off	24 Rear sprocket
2 Wheel - spoke	10 Bearing spacer	18 Lock nut - 2 off	25 Speedometer gearbox
3 Spoke set	111 Bearing - 2 off	19 Balance weight	26 Bolt - 4 off
4 Left-hand bearing carrier	12 Bearing retainer - 2 off	20 Wheel spindle	27 Washer - 8 off
5 Right-hand bearing carrier	13 Right-hand dust cover	21 Nut	28 Nut - 4 off
6 Spoke nipple	14 Left-hand dust cover	22 Washer	29 Brake disc
7 Sprocket mounting flange	15 Right-hand bearing spacer	23 Left-hand bearing spacer	30 Security bolt
8 Spacer ring	16 Chain adjuster - 2 off		

30 Disc brakes - examining and removing the brake pads

1. The brake pads of both the front and rear (where applicable) disc brakes will require renewing when the pad lining thickness reaches a minimum of 2 mm (0.08 in). It is important not to confuse this with the overall thickness of the pad plus its metal backing, since if the pad wears down to the bare metal, the disc itself will be badly damaged. Worse still, the loss in braking efficiency may give rise to a serious accident.

2 To examine or renew the pads, first detach the chrome plated cover from around the caliper; this is retained by two cross-head screws. If the two split-pins that retain the brake pads are then removed, the pads can be lifted out, one at a time.

3 When refitting the existing pads or fitting new ones, push the actuating plungers of the caliper back with a screwdriver, otherwise there will not be sufficient clearance to reinsert the pads. Use new split pins to retain the pads in place and do not omit to bend the ends of each pin over.

30.1 Split-pins retain each brake pad in position

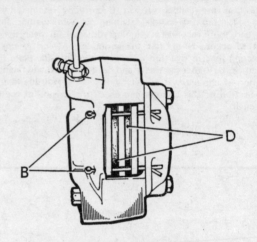

Fig. 9.41 The disc brake caliper

B Split-pins   D Brake pads

Fig. 9.42 The front brake master cylinder (rear brake master cylinder is similar in construction)

1 Pushrod
2 Piston
3 Check valve
4 Return spring
5 Primary seal
6 Circlip
7 Piston washer
8 Secondary seal
9 Spring retainer
10 Dust cover
11 Grub screw
12 Reservoir retaining nut
13 O-ring
14 Paper washer
15 Rubber riaphragm
16 Cap
17 Distance piece

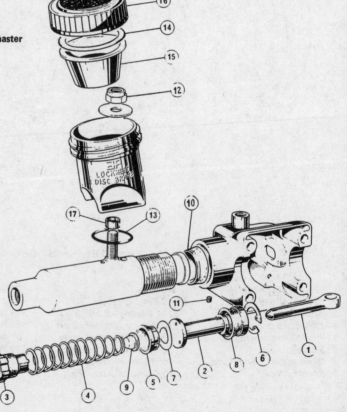

31 Rear disc brake reservoir - removal and refitting

1 The reservoir containing the fluid for the rear disc brake is located under the dualseat, behind the right-hand side panel. Being in this position, it is likely to be overlooked. It should be inspected and, if necessary, topped up, when the front brake reservoir receives attention. Full instructions for fluid replenishment are given in the Routine Maintenance Section of this Chapter.

2 To remove the reservoir, refer to the information given in Section 5 of Chapter 7 and use a similar method to drain the reservoir of fluid. Release the hydraulic hose from the master cylinder by undoing its retaining clip and pulling it from position and then gain access to the top of the reservoir by lifting the dual seat.

3 Remove the bolt which retains the reservoir assembly to the airbox, unscrew the reservoir cap with its sealing ring and lift out the rubber diaphragm to expose the head of the reservoir to

mounting piece retaining bolt. Remove this bolt with its washers and distance piece and carefully ease the reservoir away from its mounting piece whilst taking care not to tear the sealing ring sandwiched between them. Triumph recommend that this O-ring should always be renewed and this must be done if it is seen to be flattened or damaged. Flattened spring washers must also be renewed as they will have lost their locking function.

4 When reassembling the reservoir assembly and refitting it to the machine, follow a sequence which is a direct reversal of that listed above. Refer to the accompanying figure for the fitted position of any component parts and tighten the reservoir to mounting piece retaining bolt to a torque loading of 4.5 lbf ft (0.6 kgf m). Refer to Section 6 of Chapter 7 for details of bleeding the hydraulic system on completion. To bleed the system effectively the caliper must be repositioned so that the bleed screw is at the highest point of the unit. This may require some preliminary dismantling depending on the model being worked on.

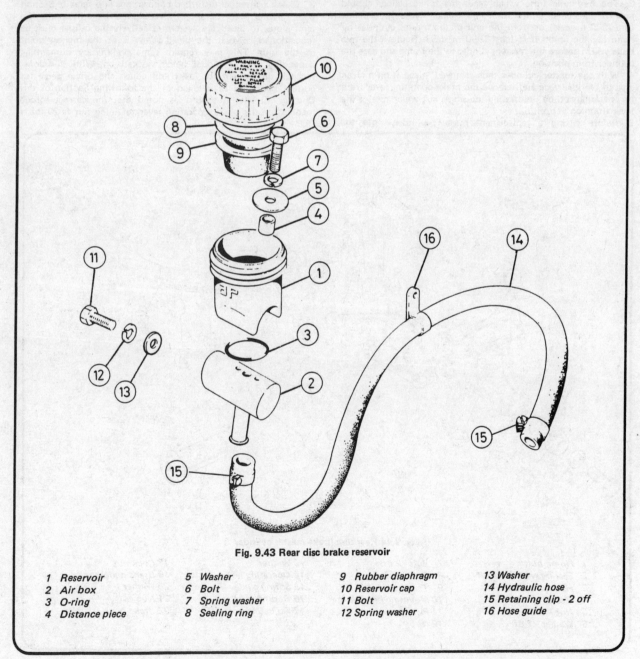

Fig. 9.43 Rear disc brake reservoir

1 Reservoir	5 Washer	9 Rubber diaphragm	13 Washer
2 Air box	6 Bolt	10 Reservoir cap	14 Hydraulic hose
3 O-ring	7 Spring washer	11 Bolt	15 Retaining clip - 2 off
4 Distance piece	8 Sealing ring	12 Spring washer	16 Hose guide

32 Rear disc brake master cylinder - removal, examination and refitting

1 The method of operation of the rear brake master cylinder is similar to that described for the front brake master cylinder in Section 5 of Chapter 7. Unlike the front cylinder however, the rear cylinder is effectively a sealed unit and in the event of failure must therefore be renewed as a complete assembly. To remove the rear cylinder, once again refer to Section 5 of Chapter 7 and drain the brake system of fluid before removing the rear wheel from the machine and proceeding as follows.

2 Detach the brake operating lever from the brake pedal by removing the nut and washer which retains it in position and then detaching the brake pedal return spring. Trace the hydraulic hose from the reservoir and detach it from the master cylinder by releasing its retaining clip and pulling it from position.

3 The hydraulic hose which feeds the brake caliper should now be detached either from the caliper hydraulic pipeline (pre-1980 models) or from the end of the master cylinder by removing the union bolt (post-1980 models). Remove the two bolts which secure the master cylinder to the frame and ease the cylinder from position.

4 With the master cylinder thus removed, place it on a clean area of work surface and remove the brake operating lever from its operating rod by unscrewing the single nut which retains the lever trunnion in position.

5 Before fitting a replacement master cylinder, note the following points. Reference to the figure accompanying this text will show that a gap of 8.9 - 9.4 mm (0.35 - 0.37 in) must exist between the face of the operating rod nut and the end of the cylinder body. Peel back the rubber boot and check this setting before fitting the master cylinder.

6 Where the caliper hose has been detached from the end of the master cylinder, check the condition of the sealing washers before refitting the union bolt. These washers must be renewed if damaged, otherwise washers which are to be reused should be annealed by heating them to a cherry red before dropping them into cold water. Take care not to overtighten the union bolt.

7 Upon refitting the brake operating lever to the brake pedal, tighten the retaining nut finger-tight only. It will be necessary to make a hook with which to pull the return spring into position over the lever, use a length of wire or string for this purpose as shown in the figure accompanying this text. With the spring properly located, tighten the lever retaining nut.

8 Check tighten all disturbed connections and refer to Section 6 of Chapter 7 for details of bleeding the brake hydraulic system. Note, to bleed the system effectively the caliper must be repositioned so that the bleed screw is at the highest point of the unit. This may require some preliminary dismantling depending on the model being worked on. With this done, recheck the system for leaks and adjust the brake pedal for correct operation as detailed in the following Section of this Chapter. Finally, it should be noted that the correct torque loading for the brake operating lever retaining nut is 20 lbf ft (2.8 kgf m).

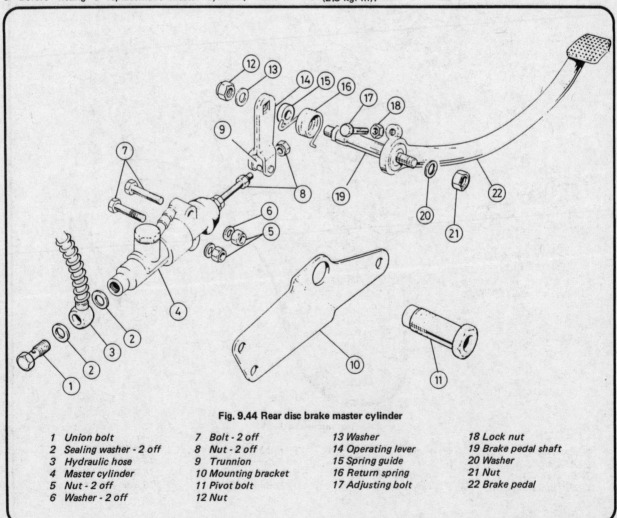

Fig. 9.44 Rear disc brake master cylinder

1 Union bolt	7 Bolt - 2 off	13 Washer
2 Sealing washer - 2 off	8 Nut - 2 off	14 Operating lever
3 Hydraulic hose	9 Trunnion	15 Spring guide
4 Master cylinder	10 Mounting bracket	16 Return spring
5 Nut - 2 off	11 Pivot bolt	17 Adjusting bolt
6 Washer - 2 off	12 Nut	

18 Lock nut
19 Brake pedal shaft
20 Washer
21 Nut
22 Brake pedal

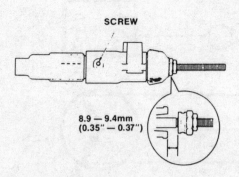

SCREW

8.9 — 9.4mm
(0.35" — 0.37")

Fig. 9.45 Master cylinder operating rod adjustment

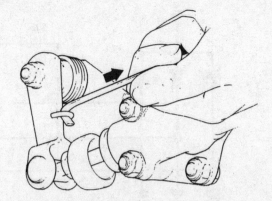

Fig. 9.46 Reconnecting the brake pedal return spring

33 Rear disc brake pedal adjustment

1 The transfer of the rear brake pedal to the right-hand side of the machine and the use of a rear disc brake has necessitated a change in the method of adjusting pedal height. As the pedal is now connected directly to the rear brake master cylinder, adjustment can be effected by slackening the nuts either side of the brake pedal arm and moving them towards the front of the machine to increase pedal height, or to the rear of the machine to reduce it.

2 Pedal height is very much a matter of personal choice. The pedal should be adjusted so that the foot can apply the brake quite naturally, without excessive ankle movement or having to raise the foot from the nearby footrest. Note, however, that it is essential that a minimum clearance of 1.6 mm (0.06 in) is maintained between the brake pedal and the footrest, see the accompanying figure. Should the brake pedal be allowed to foul the footrest, then the brake will begin to bind. Remember to adjust the stop lamp switch to suit.

3 On completion of adjustment, check-tighten the adjuster nuts and check the brake for correct operation before taking the machine onto the road. Note that the pedal should move freely about its pivot with no signs of excessive wear. Introducing oil to the pivot at the recommended service interval of 250 miles (400 km) will help keep it free of wear.

33.2 The rear brake stop lamp switch

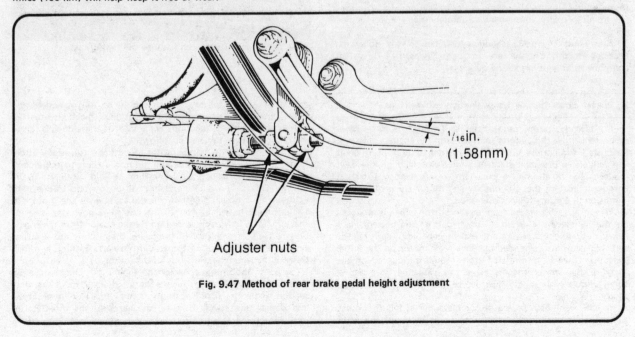

¹/₁₆in.
(1.58mm)

Adjuster nuts

Fig. 9.47 Method of rear brake pedal height adjustment

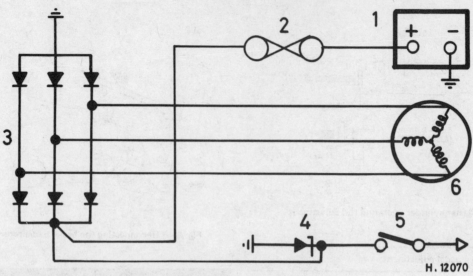

H. 12070

Fig. 9.48 The electrical charging circuit

1 Battery	3 Rectifier	5 Ignition/lighting switch
2 Fuse	4 Zener diode	6 Alternator

34 Electrical equipment - introduction

1 Reference to the Specifications Section of this Chapter will show that on later T140 and TR7 models, major changes to the electrical system have occurred. These comprise the fitting of an electric starter system to certain models, the changing of the system from positive to negative earth, the introduction of electronic ignition and a new alternator (RM 24) and the introduction of a new ignition switch and warning light console. A neutral indicator light has also been introduced and changes made in the specification and location of the switchgear.

2 Mention should also be made of the oil pressure switch, fitted into the forward facing portion of the timing cover, on the right-hand side of the machine. If the oil pressure warning lamp does not light up immediately the ignition is switched on, and it is known that the indicator bulb is not faulty, the complete switch unit must be renewed, as it cannot be repaired. It performs a vital function in giving visual indication of any serious fall in oil pressure that would ultimately cause severe engine damage.

3 Note that the test schedules detailed in the following Sections of this Chapter are for negative earth (−) system components only unless stated otherwise.

34.2 The oil pressure switch fits into the timing cover

35 Starter motor - removal, examination and refitting

1 The electric starter motor fitted to Triumph motorcycles is mounted behind the cylinder barrels, atop the engine crankcase. Since its introduction, this unit has proved to be both efficient and reliable, a desirable state of affairs especially on the TSX model where no alternative kickstart arrangement is fitted. It is most important that the battery and motor are kept in good condition to ensure efficient operation.

2 The motor is operated from a push button which is mounted on the handlebars. Current is supplied from the battery via a heavy duty solenoid switch and a cable capable of carrying the very high current demanded by the starter motor on the initial start up. Drive is transmitted from a pinion mounted on the shaft of the starter motor, via a free wheeling Borg-Warner sprag clutch with shock absorber to the intermediate timing pinion.

3 In the event of a partial or complete starter failure, do not automatically assume that the motor is at fault but check the condition of the battery, which should be fully charged, and ensure that the starter solenoid is working. Check also that all switch and wiring connections are sound. If these checks fail to effect a cure, then proceed as follows.

4 The parts of the starter motor most likely to require attention are the brushes and the commutator. Access to these components can be gained by removing the end cover from the motor. This cover is retained in position by two screws with washers. It is possible to remove the cover with the motor fitted although it will be found difficult to remove the lower of the screws. If it is decided to remove the motor from the engine unit to facilitate removal of the cover, then isolate the electrical system by disconnecting the lead from each battery terminal. Release the lead from the body of the motor.

5 Move to the timing cover of the engine unit and remove the three Allen screws which secure the small cover to its rearmost section. Note the fitted position of the longest of these screws and detach the cover to reveal the heads of the three starter motor securing screws. Remove these screws with their lock

washers, grasp the motor firmly and ease it clear of the engine unit. Note that the motor is aligned with the timing cover by means of a single dowel, pull the motor clear of this dowel before lifting it clear.

6 With the end cover removed, the brush retaining plate will be exposed. At the time of writing this Chapter, there are no spare parts listed for the starter motor and no wear limits for items such as the commutator and brushes. It is therefore advisable to remove brushes that are suspected of being overworn and to return them to a Lucas or Triumph service agent who will be able to give a professional opinion and if necessary, supply replacement items. If the motor has been removed, then ask the agent to give an opinion on the state of wear of the commutator.

7 Because both brushes will have worn together, it is advisable to renew them as a pair. Before fitting the brushes, make sure that the commutator is clean. The commutator is the copper segments on which the brushes bear. Clean the commutator with a strip of glass paper. Never use emery cloth or 'wet-and-dry' as the small abrasive fragments may embed themselves in the soft copper of the commutator and cause excessive wear of the brushes. Finish off the commutator with metal polish to give a smooth surface and finally wipe the segments over with a methylated spirits soaked rag to ensure a grease free surface. Check that the mica insulators, which lie between the segments of the commutator, are undercut. If the amount of undercut is thought to be less than ideal, then the motor should be returned to an experienced auto-electrician for recutting. As a guide, the insulation between each pair of commutator segments should lie approximately 1.0 mm (0.04 in) below the surface.

8 Note that recutting of the commutator can be undertaken at home, provided that great care is taken; new armatures are expensive. Find a broken hacksaw blade and grind the sides of the teeth so that it is the same width as the commutator grooves, then wrap some PVC tape around the other end to form a handle. Carefully re-cut the insulation whilst taking care to keep the grooves parallel and even. Do not remove an excessive amount of material.

9 Fit the brushes in their holders and check that they slide quite freely. If the original brushes are being refitted, make sure that they are fitted in their original positions as they will have worn to the profile of the commutator. The springs must act to keep the brushes firmly in contact with the commutator. If either spring has become weakened or broken, then it must be renewed.

10 If the motor has given indications of a more serious fault, the armature and field coil winding should be checked using a multimeter set on the resistance scale. To check the armature, set the meter on the ohms x 1 scale, and measure the resistance between each pair of commutator segments. In practice, the tests will identify any dead segment. Check for armature insulation faults between each segment and the metal of the armature body. An insulation failure will require the renewal of the armature. Check for continuity between the starter motor terminal and each brush lead. Infinite or zero resistance is indicative of a fault in the field coil windings.

11 Any high pitched whine or screech emitting from the motor during use, or any resistance of the armature to rotation, will indicate that the motor bearings have become seized, or are on the point of doing so. Unfortunately, these bearings are not listed as separate component parts but before going to the expense of purchasing a new motor, obtain professional advice from an experienced auto-electrician as to whether it is possible to have the bearings replaced with comparable items.

12 Any reassembly of the motor should be carried out in clean, dry conditions. Before fitting the motor to the engine unit, check that the lock washers fitted beneath the heads of the motor securing screws have not become flattened. If this is the case, then renew the washers as they will have lost their locking properties. Refitting the motor is a direct reversal of the removal procedure; remember to locate the motor over the dowel before pushing it home against the timing case.

35.4 Remove the starter motor end cover to expose the brushes

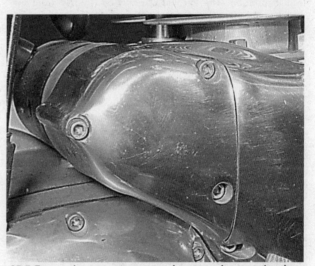
35.5 Expose the starter motor securing screws by removing the cover

36 Starter solenoid 4ST - function, removal and refitting

1 The starter solenoid is designed to work on the electro-magnetic principle. When the starter motor button is depressed, current from the battery passes through windings in the solenoid and generates an electro-magnetic force which causes a set of contact points to close. Immediately the points close, the starter motor is energised and a very heavy current is drawn from the battery.

2 This arrangement is used for two reasons. Firstly, the current drawn by the starter motor is very high, which requires the use of proportionately heavy cables to supply current from the battery to the motor. Running such heavy cables directly to the conveniently placed handlebar start switch would be cumbersome and impractical. Second, because the demands of the starter motor are so high, as short a cable run as possible is used to minimise volt drop in the circuit.

3 If the starter will not operate, first suspect a discharged battery. This can be checked by trying the horn or switching on the lights. If this check shows the battery to be in good shape, suspect the solenoid which should come into action with a pronounced click. The solenoid is located beneath the base of

the right-hand side panel and is shielded by a plastic cover which is retained in position by a single nut with plate washer; it is located close to the battery, to which it is connected by a heavy duty cable. Before condemning the starter solenoid, carry out the following test.

4 To test the operation of the solenoid, disconnect the starter motor cable at the solenoid terminal and attach multimeter probe leads to each of the solenoid terminals. Set the meter on the ohms x 1 scale. Check that the machine is in neutral gear. Switch on the ignition and press the starter button. The solenoid should operate with an audible click and the meter needle should swing across to read zero ohms (continuity). If this is not the case the solenoid can be considered defective, unless of course there is a defect in the handlebar switch and its wiring circuit.

5 To remove the solenoid, first gain proper access by detaching the right-hand side panel and then expose the solenoid by removing its cover. Isolate the electrical system by disconnecting the lead from each battery terminal and then disconnect the electrical leads from the solenoid whilst noting their fitted positions for reference when reconnecting. Detach the solenoid with its mounting bracket from the frame by removing the single bolt which retains both the bracket and engine earth lead in position. The solenoid can now be removed from the machine and placed on a clean work surface. If a replacement solenoid is to be fitted, then the mounting bracket will have to be removed from the unserviceable item and retained for the new solenoid.

6 Before fitting the replacement solenoid, check that all electrical connections and earth points are clean and free from corrosion. The procedure used for fitting a solenoid is a direct reversal of that used for removal.

36.3 The starter solenoid is covered by a plastic shield

37 Alternator RM 24 - general description

1 The RM 24 alternator is an 11 amp version of the basic 3-phase design fitted to earlier models and has been designed to physically replace the RM 21 and RM 23 units. The stator inside and outside diameters and the three fixing hole positions are identical although the lamination stack thickness has increased by 0.030 in. The RM 23 hot stacked rotor is fitted.

2 The stator has been redesigned with 9 poles which with the 6-pole rotor gives a 3-phase output. This has improved the slow speed output of the alternator. A star connection system is fitted. No maintenance is necessary, but it is advisable to check that the snap connectors in the output cable are clean and tight.

38 Alternator RM24 - removal and refitting

1 The alternator is mounted on the left-hand end of the crankshaft, behind the primary chaincase cover. Gain access to the alternator by first draining the primary chaincase of oil. Whilst the oil is being drained, release the securing bolt of the left-hand footrest and swing the footrest down to clear the chaincase. Remove the gearchange lever by releasing the pinch bolt and pulling the lever off its splined shaft. Remove the primary chaincase securing screws and nuts with washers and ease the chaincase clear of the machine whilst taking care not to tear its sealing gasket.

2 To remove the alternator stator, remove the three self-locking nuts with washers and detach the bullet connectors from the main harness. Carefully draw the alternator leads through the crankcase grommet when lifting the stator clear.

3 Remove the alternator rotor by first releasing the lock washer and then unscrewing the retaining nut. It may be necessary to lock the crankshaft in position when attempting to loosen the nut. Do this by placing the machine in gear and then applying the rear brake. The rotor is keyed to the crankshaft end and should pull clear reasonably easily.

4 Refitting the alternator is a direct reversal of the removal procedure, whilst noting the following points. Tighten the rotor securing nut to the specified torque loading and use a new lock washer to retain it in position. With the stator refitted and its retaining nuts nipped tight, check for a gap of 0.20 mm (0.008 in) between the rotor and stator at all points around the rotor circumference. With this clearance, determined, tighten the stator retaining nuts to the specified torque loading. Check the stator lead connections for cleanliness before reconnecting them and make sure that the chaincase cover gasket is serviceable before pushing the cover into position. Take note of the torque setting for the securing nuts of the cover and remember to fit a serviceable copper washer beneath each nut. On completion, refill the chaincase with the recommended quantity of oil.

39 Alternator RM24 - charging system load balancing test

1 Ensure the battery is in good condition and the terminal connections are clean and connected correctly.

2 Disconnect the battery positive terminal cable and connect a moving coil ammeter, with a scale of 0 - 25 amps, in series with the battery lead. The black ammeter lead should connect to the battery and the red lead to the harness.

3 Start the engine and increase the revs to 2000 rpm. Switch on the headlamp and select high beam. The ammeter reading should now indicate a small amount of charge. If this is not so and the reading shows a discharge, note the reading for checks in the following Sections.

4 A discharge reading indicates a possible alternator, rectifier, battery or wiring fault. Proceed to the following Section.

40 Alternator RM24 - ac output check

1 When testing the ac output of the alternator, the output between any two leads should be:

    At 1000 rpm - 4.5 volts minimum
    At 2000 rpm - 5.0 volts minimum
    At 5000 rpm - 6.5 to 7.0 volts

2 To test the alternator, disconnect the green/white, green/yellow and green/black leads that run from the alternator to the rectifier.

3 Obtain an ac voltmeter with a scale of 0 - 18 volts and connect it, with a 1 ohm resistor (100 watt rating) across the meter terminals, between the green/white and green/black alternator leads.

4 Start the engine, increase the revs to 2000 rpm and note the voltmeter reading.

5 Repeat the test with the voltmeter connected across the green/white and green/yellow leads, and again with the meter across the green/yellow and green/black leads. The voltmeter should show a reading of 5 volts minimum on all the tests. Should there be either a low or no reading on two of the tests then the alternator stator is faulty.

6 If there is a low or no reading on all of the tests then the alternator rotor has become demagnetised. They may, however, be a possible stator fault. This may be tested as shown in the following Section.

7 Should the readings all be correct, then reconnect the alternator leads and disconnect the same leads at the rectifier end. Carry out the above test again. If the test is unsatisfactory, check the leads and snap connections for deterioration. If necessary, proceed to Section 42 and test the rectifier.

41 Alternator RM 24 - stator test

1 If after carrying out the preceding test, the stator is suspected of being faulty, further investigations into its condition should be made.

2 The resistance between each of the stator leads and any other stator lead should be measured using a multi-meter set to the resistance function or a separate ohmmeter. A reading of 0.80 - 0.95 ohms should be given for any pair tested. If an infinitely high resistance (no continuity) is found, there is evidence of an open circuit in one of the stator coils. Should a low reading result, there is evidence of a short circuit in one or more coils or a short to earth. Because the specified resistance range is so small, accurate testing can only be accomplished with sensitive equipment and as such these tests should be considered as an initial investigative step. It is suggested that further testing is carried out by a qualified auto-electrician before consigning the stator to the scrap bin.

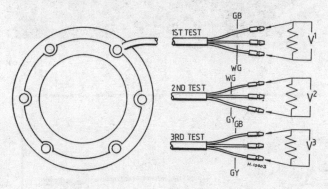

Fig. 9.49 Checking the alternator stator coils

42 Rectifier 3DS - removal and testing

1 The rectifier is located on the front of the rear mudguard behind the air filters. Access is obtained by removing the right-hand side panel.

2 **Note:** should it be necessary to remove the rectifier, never move the nut which clamps the plates together otherwise the plates will twist and the internal electrical connections will be broken. The fixing bolt and nut are both marked by circles. This indicates the No 10 x 32 UNF thread size.

3 To test the dc output of the rectifier, disconnect the dc output (brown/blue) lead and connect a dc voltmeter with a scale of 0 - 18 volts and a 1 ohm load resistor (100 watt rating) as shown in the accompanying figure. Connect the black voltmeter lead to earth and the red lead to the rectifier main positive terminal.

4 Start the engine and increase the revs to 2000 rpm. The voltmeter reading should be 9 volts minimum. **Note:** the dc circuits must not be disconnected whilst the engine is running. Should there be a low reading or no reading then the rectifier is faulty.

5 If the reading is correct and if the ammeter reading in Section 39 was showing the 'full lighting discharge, then there is no charge at all and the fault is due to an open circuit in the wiring harness.

6 If the reading is correct and the ammeter reading was only slightly below zero, then this indicates a low charge. Suspect a sulphated battery and re-test with a substitute battery. If there is no change, the fault may be due to incorrect regulating voltage. If necessary, proceed to Section 44 and test the Zener diode.

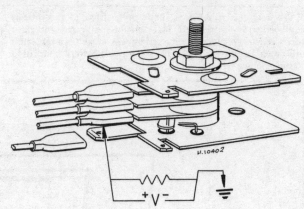

42.1 The rectifier is mounted on the rear mudguard

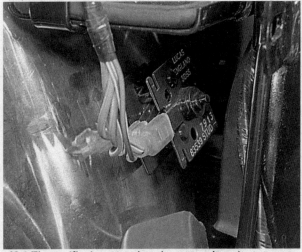

Fig. 9.50 Rectifier DC output test (negative earth system)

43 Zener diode - general description

1 The purpose of the Zener diode is to divert surplus charging current away from the battery. It therefore acts as a voltage regulator, controlling the current into the battery.

2 The body of the diode is constructed of copper to ensure maximum heat conductivity and is attached to the aluminium air box casting. This casting acts as a heatsink, dissipating excess charge current in the form of heat. As long as the diode is kept clean and the heat sink ventilated, no maintenance is necessary.

44 Zener diode ZD 715 - testing, removal and refitting

1   The following test procedure may be carried out with the diode in position on the machine. To gain access to the diode it is necessary to carry out removal of the right-hand air filter as detailed in Section 14. The diode is located at the top of the air box casting.

2   Ensure the battery is fully charged. This is essential otherwise the results of the test will not be accurate. Disconnect the cable attached to the Zener diode terminal. Refer to the accompanying figure and connect a moving coil ammeter with a minimum scale of 0 - 5 amps in series with the detached lead and the diode terminal. On negative earth systems, connect the red ammeter lead to the cable and the black lead to the terminal. On positive earth systems, connect the red lead to the terminal and the black lead to the cable.

3   Obtain a dc voltmeter with a scale of 0 - 18 volts and connect it between the diode terminal and earth. On negative earth systems, connect the red lead to the terminal. On positive earth systems, connect the black lead to the terminal.

4   Start the engine and check that all lights are switched off. Increase the engine revs until a reading of 2 amps is shown on the ammeter. At this point the voltmeter should show a reading of between 13.5 and 15.3 volts. Should the voltmeter read above or below this reading then the diode is faulty and will have to be renewed.

5   To remove the diode unscrew the retaining nut.

6   When refitting the diode, ensure that the contact surfaces of the diode and air cleaner casting are free of any contamination. This is to ensure that no air gap is allowed to exist which would reduce the heat conductivity between the two components and thus cause overheating of the diode insulation, resulting in permanent damage.

Note: the diode fixing stud nut must not be subjected to a torque of more than 2 lbf ft (27.6 kgf cm).

45 Zener diode XD 715A - testing, removal and fitting

1   To test this triple pack diode unit, fitted to the models equipped with electric start, first obtain a multimeter and set it to its resistance function or alternatively, make up a test lamp utilising a 12 volt bulb and obtain a 12 volt battery (that fitted to the machine will suffice).

2   The figure accompanying this text clearly shows the method by which each individual diode in the pack can be tested. This test should be carried out with the diode unit fitted to the machine and with its electrical leads disconnected from the main wiring of the machine; these lead connections are located beneath the carburettor assembly.

3   Test No 1 shows the positive (+) terminal of the battery connected to earth. As the test lamp is made to contact each diode lead, the bulb should be seen to illuminate thereby indicating continuity. Test No 2 shows the negative (−) terminal of the battery connected to earth. This time, the bulb of the test lamp should fail to illuminate as the lamp is connected, showing non-continuity.

4   If the tests show the diode to be unserviceable, then it must be renewed. To gain access to the diode, lift the dualseat and remove the battery. Remove the diode securing screws and remove the diode through the battery housing. Before fitting a replacement diode, check all mating surfaces and electrical connections for cleanliness.

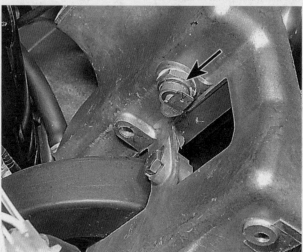

44.1 The ZD 715 Zener diode

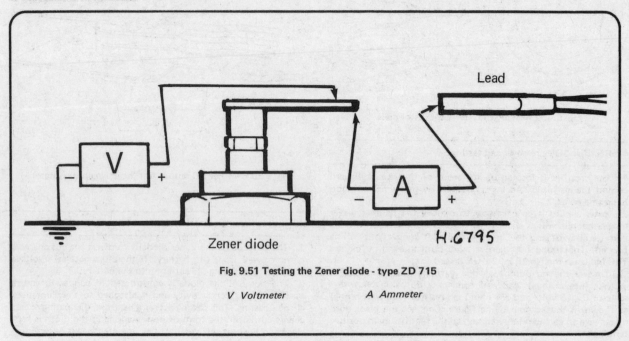

Fig. 9.51 Testing the Zener diode - type ZD 715

V Voltmeter   A Ammeter

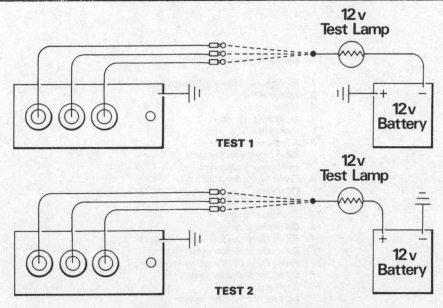

Fig. 9.52 Testing the Zener diode - type XD 715 A

46 Ignition/lighting switch 149 SA - operation

1 This switch is located in the warning light console and its key positions are as follows:

Key fully anti-clockwise — Pilot, tail and instrument lights (parking). Key may be removed.

1st position clockwise — All electrical systems off. Key may be removed.

2nd position clockwise — Ignition and electrical accessories on. Key locked in switch.

Fully clockwise — All systems (including lights) on. Key locked in switch.

On export models for the USA and Canada, with the switch in the 2nd position clockwise, the electrical accessories only are switched on. The ignition is switched on when the switch is moved to the fully clockwise position. The system is otherwise identical.

2 Note that there is no emergency start position. The machine will start with the key turned to the 'ignition' position, even with a flat battery. Refer to the following Section for details of switch removal and refitting.

47 Warning light console - bulb renewal

TR65 and TR7T models

1 The warning light console fitted to these models takes the form of a circular plate which is a press fit into what would normally be the tachometer housing. To gain access to the warning light bulbs it is first necessary to move the rubber cover clear of the ignition switch and then to unscrew the retaining ring which secures the switch to the console plate. Push the switch down so that it is free of the plate and then ease the plate clear of its housing whilst pushing the electrical wires to the warning lights further into the bottom of the housing.

2 With the bottom of the bulb holders thus exposed, each defective bulb can be pulled from its holder and renewed. Reassembly of the console is a direct reversal of the dismantling procedure. Inserting the key in the ignition switch will enable

the switch to be drawn fully up through the plate and will therefore aid refitting of its retaining ring with washer.

All other models

3 The warning light console fitted to these models is located between the speedometer and tachometer units. To gain access to the warning light bulbs, it is necessary to remove the upper half of the console and this can be done by unscrewing the three crosshead screws which secure the two halves of the console together (one screw at the front, two at the rear).

4 With the top of the console thus released, remove the rubber cover from the top of the ignition switch and unscrew the retaining ring. Remove the wave-washer. It is now possible to lift away the upper half of the console to expose the push-fit bulb retainers. The defective bulb may now be renewed.

5 Reassembly of the console is a reversal of the dismantling procedure. It is a good idea to insert the key in the ignition switch to enable it to be drawn fully up through the console; this will aid refitting of the retaining ring and washer.

47.4 Release the top of the warning light console (except TR65 and TR7T)

Wiring diagram component key

1 Right-hand handlebar switch
2 Horn switch
3 Horn
4 Headlamp dip beam
5 Dip switch]
6 Headlamp main beam
7 Main beam warning lamp
8 Headlamp flasher
9 Front brake lamp switch
10 Pilot lamp
11 Speedometer lamp
12 Tachometer lamp
13 Oil pressure switch
14 Front right-hand direction indicator
15 Direction indicator warning lamp
16 Front left-hand direction indicator
17 Direction indicator switch
18 Engine stop switch
19 Left-hand handlebar switch
20 Contact breaker
21 Ignition coil
22 Condenser
23 Lighting switch
24 Ignition switch
25 Rectifier
26 Alternator
27 Zener diode
28 Battery
29 Rear brake lamp switch
30 Direction indicator switch
31 Tail lamp
32 Stop lamp
33 Right-hand direction indicator
34 Left-hand direction indicator
35 Internal connection
36 Fuse
37 Pilot lamp switch
38 Ignition/lighting switch
39 Neutral indicator lamp
40 Neutral indicator switch
41 Transducer
42 Ignition amplifier
43 Oil pressure warning lamp
44 Headlamp switch
45 Horn/main beam flash switch

Wiring diagram colour code

B Black
K Pink
U Blue
W White
P Purple
G Green
N Brown
Y Yellow
R Red
L Light
D Dark

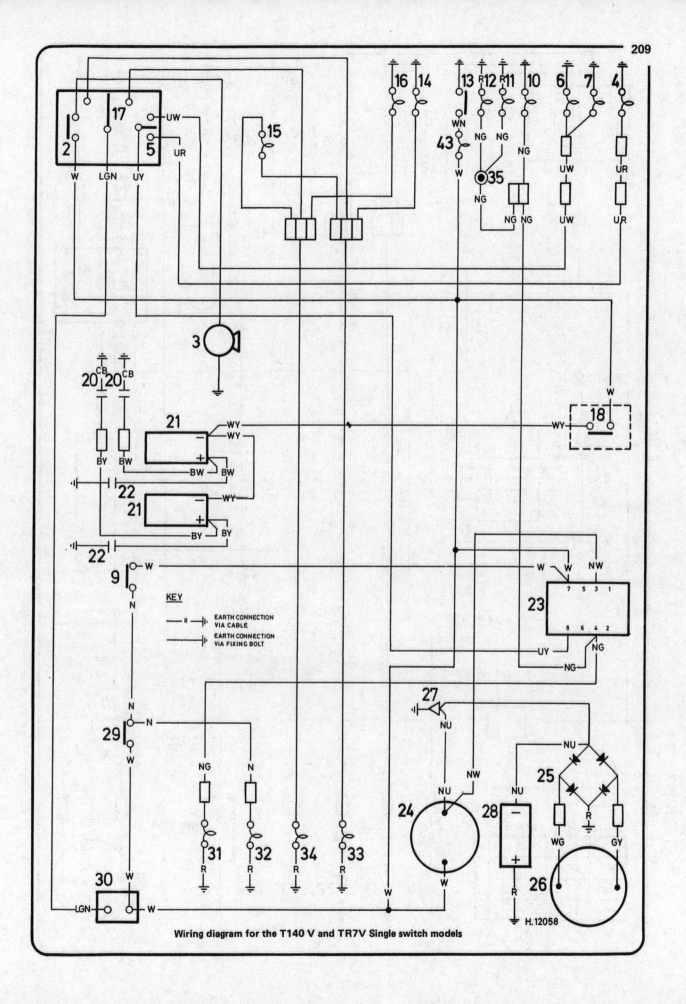

Wiring diagram for the T140 V and TR7V Single switch models

KEY

⊣R⊦ EARTH CONNECTION VIA CABLE

⊣⊦ EARTH CONNECTION VIA FIXING BOLT

H.12058

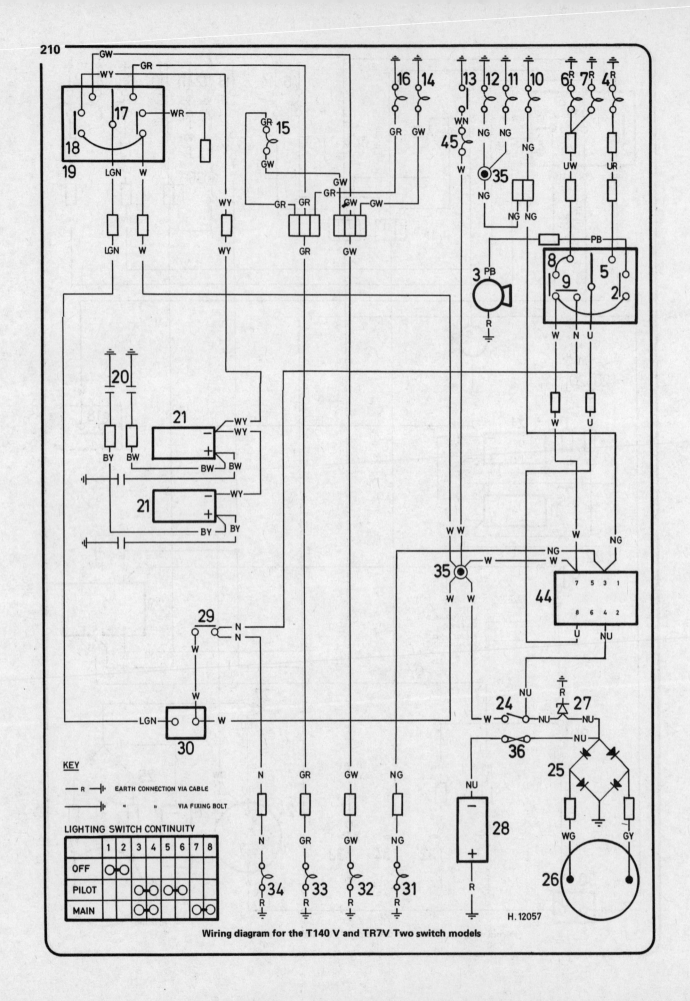

KEY

— R —||— EARTH CONNECTION VIA CABLE

——— —||— " " " VIA FIXING BOLT

LIGHTING SWITCH CONTINUITY

	1	2	3	4	5	6	7	8
OFF	O─O	O─O						
PILOT			O─O	O─O				
MAIN			O─O	O─O			O─O	

H. 12057

Wiring diagram for the T140 V and TR7V Two switch models

211

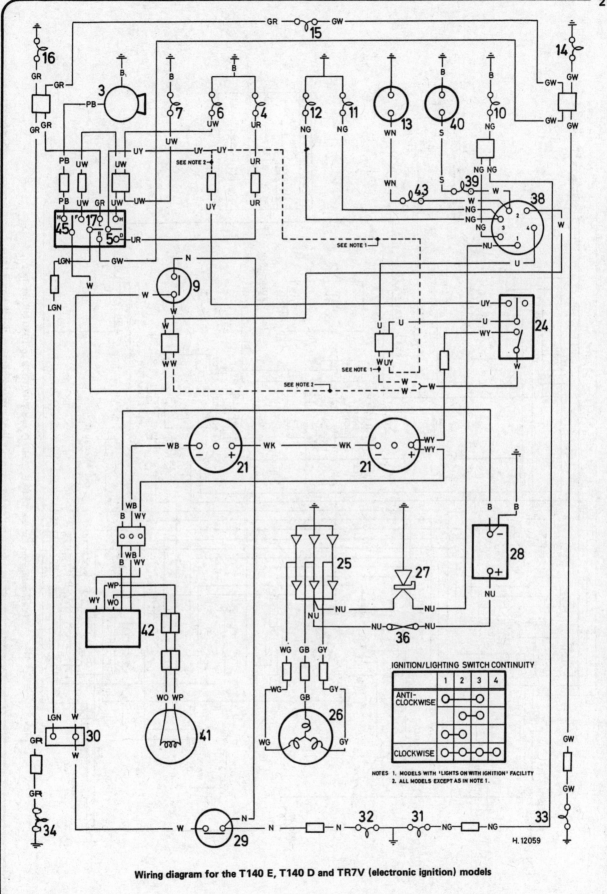

Wiring diagram for the T140 E, T140 D and TR7V (electronic ignition) models

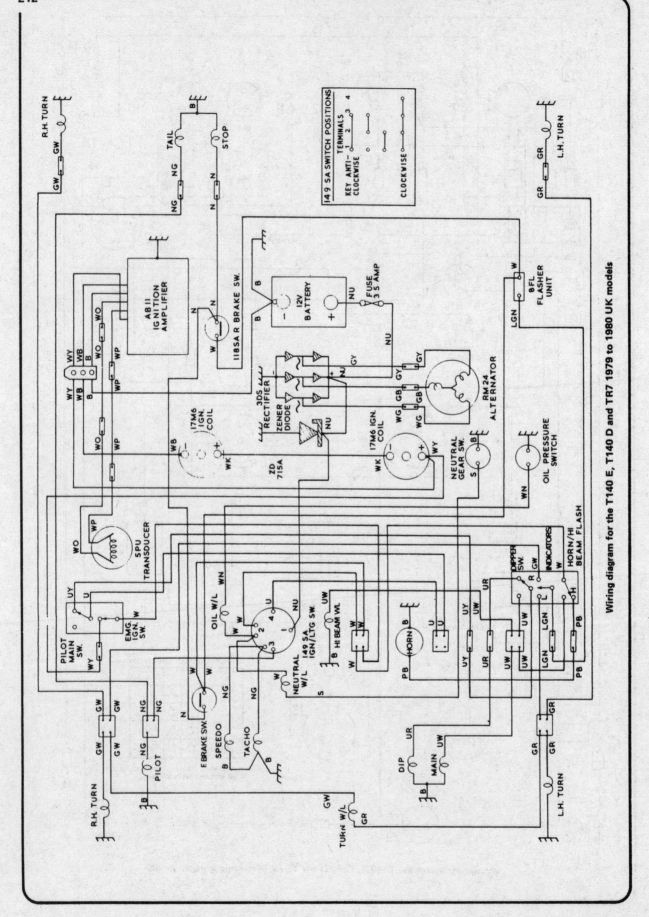

Wiring diagram for the T140 E, T140 D and TR7 1979 to 1980 UK models

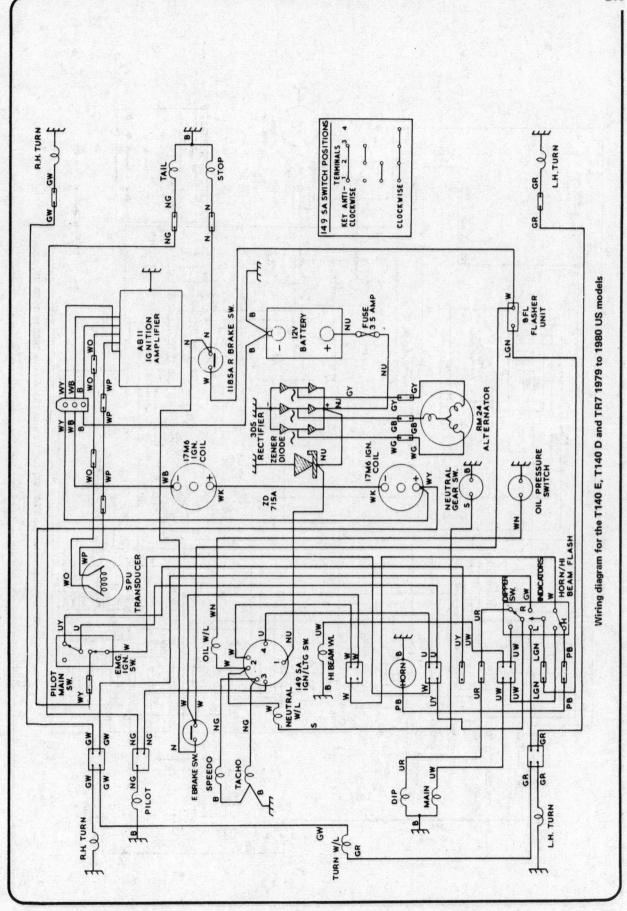

Wiring diagram for the T140 E, T140 D and TR7 1979 to 1980 US models

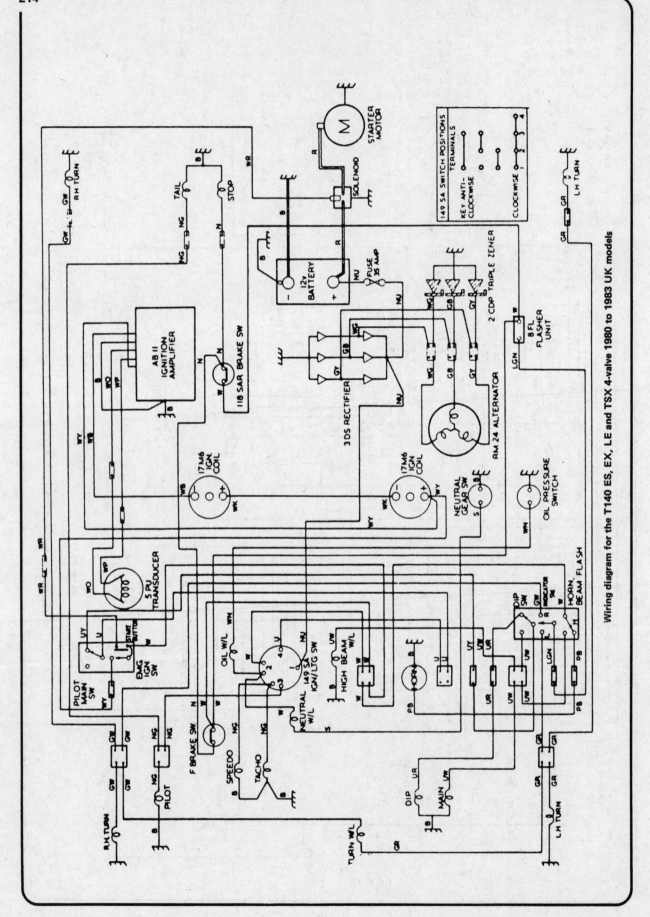

Wiring diagram for the T140 ES, EX, LE and TSX 4-valve 1980 to 1983 UK models

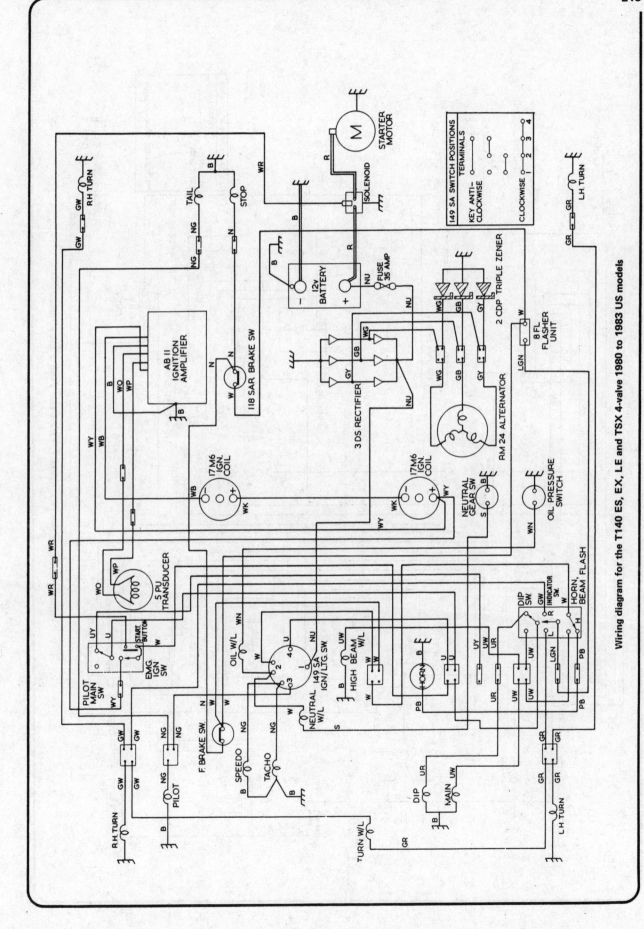

Wiring diagram for the T140 ES, EX, LE and TSX 4-valve 1980 to 1983 US models

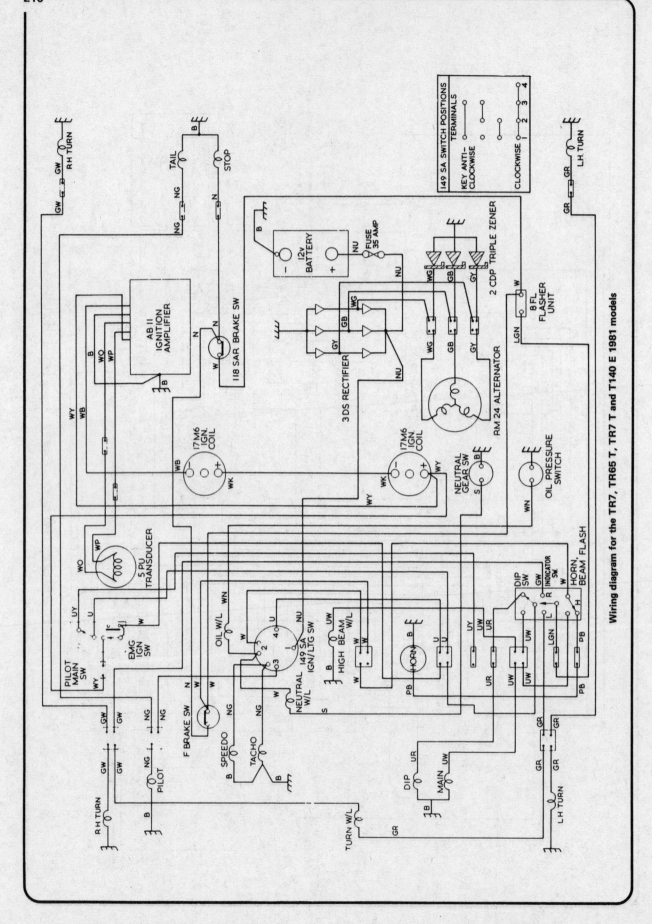

Wiring diagram for the TR7, TR65 T, TR7 T and T140 E 1981 models

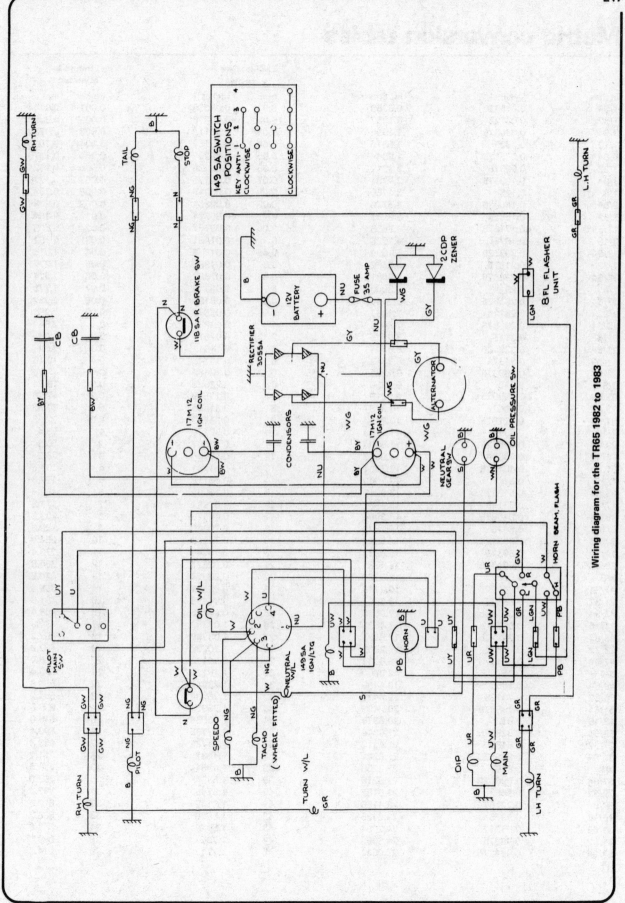

Wiring diagram for the TR65 1982 to 1983

Metric conversion tables

Inches	Decimals	Millimetres	Millimetres to Inches		Inches to Millimetres	
			mm	Inches	Inches	mm
1/64	0.015625	0.3969	0.01	0.00039	0.001	0.0254
1/32	0.03125	0.7937	0.02	0.00079	0.002	0.0508
3/64	0.046875	1.1906	0.03	0.00118	0.003	0.0762
1/16	0.0625	1.5875	0.04	0.00157	0.004	0.1016
5/64	0.078125	1.9844	0.05	0.00197	0.005	0.1270
3/32	0.09375	2.3812	0.06	0.00236	0.006	0.1524
7/64	0.109375	2.7781	0.07	0.00276	0.007	0.1778
1/8	0.125	3.1750	0.08	0.00315	0.008	0.2032
9/64	0.140625	3.5719	0.09	0.00354	0.009	0.2286
5/32	0.15625	3.9687	0.1	0.00394	0.01	0.254
11/64	0.171875	4.3656	0.2	0.00787	0.02	0.508
3/16	0.1875	4.7625	0.3	0.01181	0.03	0.762
13/64	0.203125	5.1594	0.4	0.01575	0.04	1.016
7/32	0.21875	5.5562	0.5	0.01969	0.05	1.270
15/64	0.234375	5.9531	0.6	0.02362	0.06	1.524
1/4	0.25	6.3500	0.7	0.02756	0.07	1.778
17/64	0.265625	6.7469	0.8	0.03150	0.08	2.032
9/32	0.28125	7.1437	0.9	0.03543	0.09	2.286
19/64	0.296875	7.5406	1	0.03937	0.1	2.54
5/16	0.3125	7.9375	2	0.07874	0.2	5.08
21/64	0.328125	8.3344	3	0.11811	0.3	7.62
11/32	0.34375	8.7312	4	0.15748	0.4	10.16
23/64	0.359375	9.1281	5	0.19685	0.5	12.70
3/8	0.375	9.5250	6	0.23622	0.6	15.24
25/64	0.390625	9.9219	7	0.27559	0.7	17.78
13/32	0.40625	10.3187	8	0.31496	0.8	20.32
27/64	0.421875	10.7156	9	0.35433	0.9	22.86
7/16	0.4375	11.1125	10	0.39370	1	25.4
29/64	0.453125	11.5094	11	0.43307	2	50.8
15/32	0.46875	11.9062	12	0.47244	3	76.2
31/64	0.484375	12.3031	13	0.51181	4	101.6
1/2	0.5	12.7000	14	0.55118	5	127.0
33/64	0.515625	13.0969	15	0.59055	6	152.4
17/32	0.53125	13.4937	16	0.62992	7	177.8
35/64	0.546875	13.8906	17	0.66929	8	203.2
9/16	0.5625	14.2875	18	0.70866	9	228.6
37/64	0.578125	14.6844	19	0.74803	10	254.0
19/32	0.59375	15.0812	20	0.78740	11	279.4
39/64	0.609375	15.4781	21	0.82677	12	304.8
5/8	0.625	15.8750	22	0.86614	13	330.2
41/64	0.640625	16.2719	23	0.09551	14	355.6
21/32	0.65625	16.6687	24	0.94488	15	381.0
43/64	0.671875	17.0656	25	0.98425	16	406.4
11/16	0.6875	17.4625	26	1.02362	17	431.8
45/64	0.703125	17.8594	27	1.06299	18	457.2
23/32	0.71875	18.2562	28	1.10236	19	482.6
47/64	0.734375	18.6531	29	1.14173	20	508.0
3/4	0.75	19.0500	30	1.18110	21	533.4
49/64	0.765625	19.4469	31	1.22047	22	558.8
25/32	0.78125	19.8437	32	1.25984	23	584.2
51/64	0.796875	20.2406	33	1.29921	24	609.6
13/16	0.8125	20.6375	34	1.33858	25	635.0
53/64	0.828125	21.0344	35	1.37795	26	660.4
27/32	0.84375	21.4312	36	1.41732	27	685.8
55/64	0.859375	21.8281	37	1.4567	28	711.2
7/8	0.875	22.2250	38	1.4961	29	736.6
57/64	0.890625	22.6219	39	1.5354	30	762.0
29/32	0.90625	23.0187	40	1.5748	31	787.4
59/64	0.921875	23.4156	41	1.6142	32	812.8
15/16	0.9375	23.8125	42	1.6535	33	838.2
61/64	0.953125	24.2094	43	1.6929	34	863.6
31/32	0.96875	24.6062	44	1.7323	35	889.0
63/64	0.984375	25.0031	45	1.7717	36	914.4

Conversion factors

Length (distance)

	X				X		
Inches (in)	X	25.4	= Millimetres (mm)		X	0.0394	= Inches (in)
Feet (ft)	X	0.305	= Metres (m)		X	3.281	= Feet (ft)
Miles	X	1.609	= Kilometres (km)		X	0.621	= Miles

Volume (capacity)

	X				X		
Cubic inches (cu in; in^3)	X	16.387	= Cubic centimetres (cc; cm^3)		X	0.061	= Cubic inches (cu in; in^3)
Imperial pints (Imp pt)	X	0.568	= Litres (l)		X	1.76	= Imperial pints (Imp pt)
Imperial quarts (Imp qt)	X	1.137	= Litres (l)		X	0.88	= Imperial quarts (Imp qt)
Imperial quarts (Imp qt)	X	1.201	= US quarts (US qt)		X	0.833	= Imperial quarts (Imp qt)
US quarts (US qt)	X	0.946	= Litres (l)		X	1.057	= US quarts (US qt)
Imperial gallons (Imp gal)	X	4.546	= Litres (l)		X	0.22	= Imperial gallons (Imp gal)
Imperial gallons (Imp gal)	X	1.201	= US gallons (US gal)		X	0.833	= Imperial gallons (Imp gal)
US gallons (US gal)	X	3.785	= Litres (l)		X	0.264	= US gallons (US gal)

Mass (weight)

	X				X		
Ounces (oz)	X	28.35	= Grams (g)		X	0.035	= Ounces (oz)
Pounds (lb)	X	0.454	= Kilograms (kg)		X	2.205	= Pounds (lb)

Force

	X				X		
Ounces-force (ozf; oz)	X	0.278	= Newtons (N)		X	3.6	= Ounces-force (ozf; oz)
Pounds-force (lbf; lb)	X	4.448	= Newtons (N)		X	0.225	= Pounds-force (lbf; lb)
Newtons (N)	X	0.1	= Kilograms-force (kgf; kg)		X	9.81	= Newtons (N)

Pressure

	X				X		
Pounds-force per square inch (psi; lbf/in^2; lb/in^2)	X	0.070	= Kilograms-force per square centimetre (kgf/cm^2; kg/cm^2)		X	14.223	= Pounds-force per square inch (psi; lbf/in^2; lb/in^2)
Pounds-force per square inch (psi; lbf/in^2; lb/in^2)	X	0.068	= Atmospheres (atm)		X	14.696	= Pounds-force per square inch (psi; lbf/in^2; lb/in^2)
Pounds-force per square inch (psi; lbf/in^2; lb/in^2)	X	0.069	= Bars		X	14.5	= Pounds-force per square inch (psi; lbf/in^2; lb/in^2)
Pounds-force per square inch (psi; lbf/in^2; lb/in^2)	X	6.895	= Kilopascals (kPa)		X	0.145	= Pounds-force per square inch (psi; lbf/in^2; lb/in^2)
Kilopascals (kPa)	X	0.01	= Kilograms-force per square centimetre (kgf/cm^2; kg/cm^2)		X	98.1	= Kilopascals (kPa)
Millibar (mbar)	X	100	= Pascals (Pa)		X	0.01	= Millibar (mbar)
Millibar (mbar)	X	0.0145	= Pounds-force per square inch (psi; lbf/in^2; lb/in^2)		X	68.947	= Millibar (mbar)
Millibar (mbar)	X	0.75	= Millimetres of mercury (mmHg)		X	1.333	= Millibar (mbar)
Millibar (mbar)	X	0.401	= Inches of water (inH$_2$O)		X	2.491	= Millibar (mbar)
Millimetres of mercury (mmHg)	X	0.535	= Inches of water (inH$_2$O)		X	1.868	= Millimetres of mercury (mmHg)
Inches of water (inH$_2$O)	X	0.036	= Pounds-force per square inch (psi; lbf/in^2; lb/in^2)		X	27.68	= Inches of water (inH$_2$O)

Torque (moment of force)

	X				X		
Pounds-force inches (lbf in; l b in)	X	1.152	= Kilograms-force centimetre (kgf cm; kg cm)		X	0.868	= Pounds-force inches (lbf in; lb in)
Pounds-force inches (lbf in; l b in)	X	0.113	= Newton metres (Nm)		X	8.85	= Pounds-force inches (lbf in; lb in)
Pounds-force inches (lbf in; l b in)	X	0.083	= Pounds-force feet (lbf ft; lb ft)		X	12	= Pounds-force inches (lbf in; lb in)
Pounds-force feet (lbf ft; lb ft)	X	0.138	= Kilograms-force metres (kgf m; kg m)		X	7.233	= Pounds-force feet (lbf ft; lb ft)
Pounds-force feet (lbf ft; lb ft)	X	1.356	= Newton metres (Nm)		X	0.738	= Pounds-force feet (lbf ft; lb ft)
Newton metres (Nm)	X	0.102	= Kilograms-force metres (kgf m; kg m)		X	9.804	= Newton metres (Nm)

Power

	X				X		
Horsepower (hp)	X	745.7	= Watts (W)		X	0.0013	= Horsepower (hp)

Velocity (speed)

	X				X		
Miles per hour (miles/hr; mph)	X	1.609	= Kilometres per hour (km/hr; kph)		X	0.621	= Miles per hour (miles/hr; mph)

Fuel consumption

	X				X		
Miles per gallon, Imperial (mpg)	X	0.354	= Kilometres per litre (km/l)		X	2.825	= Miles per gallon, Imperial (mpg)
Miles per gallon, US (mpg)	X	0.425	= Kilometres per litre (km/l)		X	2.352	= Miles per gallon, US (mpg)

Temperature

Degrees Fahrenheit = (°C x 1.8) + 32   Degrees Celsius (Degrees Centigrade; °C) = (°F – 32) x 0.56

It is common practice to convert from miles per gallon (mpg) to litres/100 kilometres (l/100km), where mpg (Imperial) x l/100 km = 282 and mpg (US) x l/100 km = 235

English/American terminology

Because this book has been written in England, British English component names, phrases and spellings have been used throughout. American English usage is quite often different and whereas normally no confusion should occur, a list of equivalent terminology is given below.

English	American	English	American
Air filter	Air cleaner	Number plate	License plate
Alignment (headlamp)	Aim	Output or layshaft	Countershaft
Allen screw/key	Socket screw/wrench	Panniers	Side cases
Anticlockwise	Counterclockwise	Paraffin	Kerosene
Bottom/top gear	Low/high gear	Petrol	Gasoline
Bottom/top yoke	Bottom/top triple clamp	Petrol/fuel tank	Gas tank
Bush	Bushing	Pinking	Pinging
Carburettor	Carburetor	Rear suspension unit	Rear shock absorber
Catch	Latch	Rocker cover	Valve cover
Circlip	Snap ring	Selector	Shifter
Clutch drum	Clutch housing	Self-locking pliers	Vise-grips
Dip switch	Dimmer switch	Side or parking lamp	Parking or auxiliary light
Disulphide	Disulfide	Side or prop stand	Kick stand
Dynamo	DC generator	Silencer	Muffler
Earth	Ground	Spanner	Wrench
End float	End play	Split pin	Cotter pin
Engineer's blue	Machinist's dye	Stanchion	Tube
Exhaust pipe	Header	Sulphuric	Sulfuric
Fault diagnosis	Trouble shooting	Sump	Oil pan
Float chamber	Float bowl	Swinging arm	Swingarm
Footrest	Footpeg	Tab washer	Lock washer
Fuel/petrol tap	Petcock	Top box	Trunk
Gaiter	Boot	Torch	Flashlight
Gearbox	Transmission	Two/four stroke	Two/four cycle
Gearchange	Shift	Tyre	Tire
Gudgeon pin	Wrist/piston pin	Valve collar	Valve retainer
Indicator	Turn signal	Valve collets	Valve cotters
Inlet	Intake	Vice	Vise
Input shaft or mainshaft	Mainshaft	Wheel spindle	Axle
Kickstart	Kickstarter	White spirit	Stoddard solvent
Lower leg	Slider	Windscreen	Windshield
Mudguard	Fender		

Index